108th International Scientific and Technical Conference of the Association of Automotive Engineers 2019

Intelligent Car Systems: Development, Research, Certification

IOP Conference Series: Materials Science and Engineering
Volume 695

Nizhny Novgorod, Russia
25 - 26 September 2019

ISBN: 978-1-7138-0987-6
ISSN: 1757-8981

Printed from e-media with permission by:

Curran Associates, Inc.
57 Morehouse Lane
Red Hook, NY 12571

proceedings.com

Some format issues inherent in the e-media version may also appear in this print version.

Institute of Physics
Dirac House, Temple Back
Bristol BS1 6BE UK

Phone: 44 1 17 929 7481
Fax: 44 1 17 920 0979

techtracking@iop.org

Additional copies of this publication are available from:

Curran Associates, Inc.
57 Morehouse Lane
Red Hook, NY 12571 USA
Phone: 845-758-0400
Fax: 845-758-2633
Email: curran@proceedings.com
Web: www.proceedings.com

TABLE OF CONTENTS

USE OF ARTICULATED TRANSPORT SYSTEMS IN THE MINING INDUSTRY 1
E E Bazhenov, D O Chernyshev, L V Bazhenova

EXPERIMENTAL RESEARCH OF THE ELECTROMECHANICAL STEERING SYSTEM 6
A V Tumasov, D Yu Tyugin, P O Beresnev, D M Porubov, V I Filatov

RESEARCHING ROLLING RESISTANCE OF THE WHEELED FORESTRY TRACTOR AT SKIDDING 16
Vladislav Klubnichkin, Evgeny Klubnichkin, Vladimir Makarov, Maxim Podrubalov

RESEARCHING IMPACT OF PRESSURE RE-DISTRIBUTION ALONG THE LENGTH OF THE TRACK GROUND CONTACTING AREA ON RESISTANCE OF THE TIMBER-HAULING VEHICLES' TRAVEL OVER SNOW 23
Evgeny Klubnichkin, Vladislav Klubnichkin, Alexander Erkhov, Alexey Alyabiev

OPTIMIZATION OF THE INTERNAL DAMPING RUBBER ELEMENTS OF THE ROLLER WHEEL OF TRACKED SYSTEM 34
S A Korostelev, A F Verbilov

AUTONOMOUS CONTROL SYSTEM PROTOTYPE FOR GAZ A65R32 41
M. I. Tiurikov, N. V. Malinin, T. S. Evdokimova, A. A. Sinodkin, L. O. Fedosova

CALCULATION AND EXPERIMENTAL RESEARCH OF LIQUID FLOW THROUGH A RADIATOR OF THE ELECTRIC CAR POWER BATTERY THERMAL MANAGMENT SYSTEM 46
A.M. Meshkov, A.A. Barinov, A.Y. Sizov, T.G. Skrebneva, I.A. Korotunova

HARDWARE IMPLEMENTATION OF THE MONITORING SYSTEM AND CONTROL UNITS OF THE ROBOTIC VEHICLE "MODULAR CONTROLLER" 54
M.A. Kolobov, D.V. Bushmanov, A.V. Okunev, A.V. Kocherov

DEVELOPMENT AND IMPLEMENTATION OF A SOFTWARE AND HARDWARE COMPLEX OF THE CONTROLS AUTOMATION INTO THE SELF-DRIVING CAR SYSTEM BASED ON THE "GAZELLE NEXT" VEHICLE 62
M M Voronin, P A Anisimov, A V Zolotov

PREDICTION OF A RESIDUAL OPERATING LIFE OF ENGINES 68
Y Bazhenov, M Bazhenov

ASSESSMENT OF THE TECHNICAL CONDITION OF THE UNITS OF MOBILE ROBOTECHNICAL COMPLEXES BASED ON THE APPROACHES OF NONLINEAR DYNAMICS AND ARTIFICIAL INTELLIGENCE 72
Yu.G. Kabaldin, D. A. Shatagin, D. A. Sidorenkov, M.V. Zhelonkin, M. S. Anosov

UNMANNED VEHICLE LIDAR-CAMERA SENSOR FUSION TECHNOLOGY 77
V.D. Kuzin, P.I. Pronin, D.N. Zarubin, K.M. Shashkina

INVESTIGATION OF COLLISION WARNING POSSIBILITIES BY MEANS OF GNSS RECEIVERS OF ANDROID SMARTPHONES 82
Roman Kulikov, Alexander Chugunov, Vladimir Zamolodchikov

APPLICATION OF SYSTEMS OF TELEMATICS ON COMMERCIAL VEHICLES 89
H M Saleh, R V Nuzhdin, S V Kurochkin

MULTICRITERIA OPTIMIZATION IN THE STUDY OF ACTIVE VEHICLE SAFETY 96
A.I. Markovnina, V.S. Makarov, V.V. Belyakov

MONITORING SYSTEMS DEVELOPMENT USING SITUATIONAL APPROACH AND FUZZY LOGIC 102
A.E. Ermilov

DESIGNED GEARBOX RELIABILITY ASSESSMENT TECHNIQUES BASED ON THE REAL ONE 109
R.R. Yusupov, A.D. Yarzhemsky, A.V. Alexeyev, M.G. Korchazhkin, L.A. Berdnikov

DESIGNED BRAKE SYSTEM RELIABILITY ASSESSMENT TECHNIQUES APPLICATION ON THE BASIS OF THE REAL ONE 115
R.R. Yusupov, A.V. Alexeev, A.D. Yarzhemsky, M.I. Chichkina, L.A. Berdnikov

APPLICATION OF ASSESSMENT TECHNIQUES OF RELIABILITY OF THE DESIGNED STEERING SYSTEM BASED ON THE REAL ONE 122
R.R. Yusupov, A.V. Alexeev, A.D. Yarzhemsky, M.G. Korchazhkin, A.S. Brovkin

BEARING FLOATATION OF FOREST MACHINES (THEORETICAL CALCULATION) 127
E G Khitrov, A V Andronov

COMPARISON OF THE FOREST MACHINE MOVER AVERAGE AND NOMINAL PRESSURE ON THE SOIL 133
E G Khitrov, A V Andronov

ENGINE-AND-TRANSMISSION INSTALLATION FREQUENCY ANALYSIS OF A WHEELED FORESTRY TRACTOR 138
A.V. Churakov, V.A. Sokolova, V.A. Markov, G.K. Parfenopulo, V.A. Kalyashov

COMPOSITION RESEARCH OF GENERATOR GAS FOR APPLICATION AS MOTOR FUEL 142
S.A. Plotnikov, A.L. Biryukov, A.S- Zubakin

NEURAL NETWORKS MODIFICATION FOR SOLVING THE TRAFFIC SIGNS DETECTION PROBLEM 149
A.V. Devyatkin, D. M. Filatov

CHOICE OF METAL MATERIALS FOR RESPONSIBLE PARTS AND VEHICLE ASSEMBLIES OPERATING UNDER THE CONDITIONS OF LOW TEMPERATURES 155
Y.G. Kabaldin, A.A. Khlybov, M.S. Anosov, D.A. Shatagin, A.A. Golovin, M.D. Scriabin

DEVELOPMENT OF DESIGN PROJECT OF MINI CROSS-COUNTRY VEHICLE KORSAK 161
A.G. Kapustin, A.V. Papunin, V.S. Makarov, V.V. Belyakov, M.E. Gnenik

SOUND GENERATION UNDER ROTARY-SCREW PROPULSION UNIT BASE CYLINDER BENDING 166
I. A. Erasov, I.G. Kuklina, Y.I. Molev, D.S. Mokerov, T.G. Skrebneva

IMPROVEMENT OF CAR REPAIRABILITY BASED ON THE MATERIAL DESTRUCTION ASSESSMENT DURING ITS OPERATION 171
N.A. Kuzmin, D.A. Goncharova, G.V. Pachurin, A.A. Filippov, I.A. Korotunova

OPTIMIZATION OF PERLITE STEELS MECHANICAL PROPERTIES FOR CAR FASTENERS STAMPING 177
A.N. Kuzmin, A.A. Filippov, G.V. Pachurin, D.A. Goncharova

EVALUATION OF MEDIUM FACTOR INFLUENCE ON CAR MATERIALS FATIGUE RESISTANCE 183
D.A. Goncharova, G.V. Pachurin, A.A. Filippov, I.G. Trunova

THE CONSIDERATION OF ROAD LONGITUDINAL SLOPES IN AUTOMOBILE FUEL CONSUMPTION RATIONING 190
G.V. Borisov, N.A. Kuzmin, J.I. Molev, I.R. Mavleev, I.I. Salakhov

ON COMPARING RADAR PERFORMANCE AND CONDUCTING THEIR ASSESSMENT TESTS 194
A. V. Myakinkov, R. S. Fadeev, V. N. Burov, A.D. Yarjemsky, P. S. Rogov, Y. P. Trusov, E.I. Toropov, P. S. Moshkov, D.A. Kolesnitcenko

Author Index

IOP Conf. Series: Materials Science and Engineering **695** (2019) 011001 doi:10.1088/1757-899X/695/1/011001

The Automotive Engineers Association (AAE) is an independent public association which was formed in September 1991. The necessity of such institution was determined by the change in the management structure of the economy of the country in general and the automotive industry in particular. The branch institutes, that generally carried out scientific and technical studies, were weakened or lost. It led to a great need of an organization which provided an opportunity for specialists of various levels and areas to communicate with each other and engage in work together.

In October 1990, a constituent conference was held with representatives of 30 organizations, including automobile factories, research institutes, higher education institutions, etc. The Association's Board and the President were elected in that day. On September 25, 1991, the AAE was registered with the Ministry of Justice of the USSR as an all-Union public organization. In 1992, the AAE was admitted to the International Federation of Automotive Engineering Societies - FISITA. Nowadays, the AAE is a member of the Chamber of Commerce and Industry of the Russian Federation (CCI RF) and the European Automotive Engineers Community (EAEC). On September 30, 1993, the Association was re-registered with the Ministry of Justice of the Russian Federation.

The 108eth International Technical Scientific Conference of Automotive Engineers Association (AAE) "Intelligent automotive systems: development, research, certification" took place in Nizhny Novgorod State Technical University (NNSTU), Russia, September 25-26, 2019. The main objective of the conference was knowledge exchange, discussion of problems and inspiring cooperation in development, research, production and operation of modern vehicles with intelligent drivers assistant system.
More than 200 people from Russia and abroad visited the conference during those two days. Also many scientists and PhD students from leading Russian universities were actively

IOP Conf. Series: Materials Science and Engineering **695** (2019) 011001 doi:10.1088/1757-899X/695/1/011001

engaged in the event. Topics and questions discussed during the conference appeared pressing and needed further research and discussion not only as a part of AAE events but also other conferences, symposiums and forums related to research and development of ADAS and autonomous driving vehicles.

Peer review statement

All papers published in this volume of *IOP Conference Series: Materials Science and Engineering* have been peer reviewed through processes administered by the proceedings Editors. Reviews were conducted by expert referees to the professional and scientific standards expected of a proceedings journal published by IOP Publishing.

USE OF ARTICULATED TRANSPORT SYSTEMS IN THE MINING INDUSTRY

E E Bazhenov[1], D O Chernyshev[2,4] and L V Bazhenova[3]

[1] Bauman Moscow State Technical University, Baumanskaya second street, 5/1, Moscow, Russia
[2] Ural State Forest Engineering University, Siberian highway 37, Ekaterinburg, Russia
[3] Ural State University of Railway Transport, Kolmogorova street, 66, Ekaterinburg, Russia
[4] den_is-best@mail.ru

Abstract. The work is devoted to the possibility and prospects of the use of all-wheels drive articulated transport systems in the mining complex. A comparative analysis of the traditional methods of exporting minerals in open pit mining and the method of using active trailed elements are given. The trailer has a load factor several times higher than the same rate for mining dump trucks. The use of an active trailer makes it possible to reduce the mass of the tractor and trailer by almost 40 tons and increase the specific power of the road train.

1. Introduction

The use of all-wheel drive transport systems based on the activation of the trailer and the use of articulated transport and technological machines is one of the promising directions in solving many problems arising from the operation of transport and technological complexes in winter roads, dirt roads during the muddy season and in other specific conditions [3]

The concept of “articulated transport system” (ATS) is very broad and includes a whole range of machines that have at least two sections connected by a hinge with one or more degrees of freedom. As a special, but rather widespread, case of ATS, road trains with active trailers or active ATS (AATS) should be considered.

The use of AATS and ATS in the sectors of the mining complex, agriculture, as well as in the sectors of the forest, oil and gas complexes and other raw material sectors of the economy makes it possible to create a wide range of technological and transport systems.

The problem of improving the operational properties of transport systems is one of the main in mechanical engineering, the solution of which should be carried out in various directions: increasing productivity, improving economic and environmental indicators, increasing operational reliability, improving and automating transport control systems, and a whole range of other theoretical and experimental focus [1,2]

2. Theoretical part

Let us analyze traditional transport systems operating in open pit mining. As a rule, BelAZ family dump trucks are used as a transport unit. As the main comparative characteristics, we take the following:

- Q_v – carrying capacity, kN;
- G_v – own weight of the vehicle, kN;
- N – engine power, кВт;
- k – factor carrying capacity.

$$k = \frac{Q_v}{G_v} \tag{1}$$

Table 1. Comparative characteristics of dump trucks

Car model and make	Q_v, kN	G_v, kN	N, kW	k
BelAZ - 548A	400	288	367,7	1,4
BelAZ – 549	750	669,8	772	1,2
BelAZ - 75214	1800	1570	1 691	1,15
MAZ- 5335	80	67,25	132,4	1,19
KamAZ – 55102	70	90,75	154	0,77
MAZ - 75165	245	223,5	346	1,1

An analysis of the technical characteristics in table 1 shows that BelAZ dump trucks are no exception among other trucks in terms of load capacity.

AATS means at least two sections, one of which carries a technological function. The trailer link in the composition of the road train is a section that carries a technological function [4,5,6]

Consider the characteristics of heavy-duty trailers manufactured by the industry. Options for comparison:

- make and model of the base tractor;
- Q_{Tr} – carrying capacity trailer, kN;
- G_{Tr} – own weight trailer, kN;
- k_{Tr} – factor carrying capacity trailer;
- N – base tractor engine power, kW;
- Q_T – carrying capacity tractor, kN;
- G_T – total base tractor weight, kN;
- k_{at} – ratio of full load trailer weight to full tractor weight.

The coefficient of the load capacity of the trailer is determined similarly to the coefficient of the load capacity of dump trucks. To compare the ratio of the total weight of the trailer and the tractor, we introduce the concept of the coefficient of the total mass of the road train k_{at}. The coefficient of the total mass of the road train will be determined as the ratio of the total weight of the trailer to the full weight of the tractor:

$$k_{at} = \frac{G_{Tr} + Q_{Tr}}{G_T + Q_T} \quad (2)$$

The results of the analysis are shown in table 2.

Table 2. Comparative characteristics of the trailer

Trailer brand	Q_{Tr}, kN	G_{Tr}, kN	k_{Tr}	Base tractor	N, kW	G_T, kN	k_{at}
MAZ – 8378	145	55	2,64	MAZ – 5335	132,4	147,3	1,36
CMZAP-5523	250	67,5	3,7	KrAZ - 258	176,5	192	1,65
CMZAP -5208	400	109	3,67	MAZ – 537P	525	440	1,16
CMZAP -5212	600	139	4,32	MAZ – 537P	525	440	1,68

An increase in the carrying capacity of a trailer increases its carrying capacity. This is due to an increase in the number of wheels.

As can be seen from the data presented, regardless of the carrying capacity of the trailer, the coefficient of its carrying capacity and the type of tractor, the value of the coefficient of the total mass of road trains is in the range 1.3 - 1.7. The reason is that universal vehicles are used as tractors, designed both for transporting goods on a loading platform (cargo weight is considered in G_T), and for towing trailers.

IOP Conf. Series: Materials Science and Engineering **695** (2019) 012001 doi:10.1088/1757-899X/695/1/012001

3. Experimental part

If you determine the necessary weight of the traction link for the trailer CMZAP-5212 according to the conditions of adhesion of the wheels to the ground, then it turns out to be 23.4 kN, which corresponds to its own weight MAZ - 537P.

Let us consider the possibility of creating an articulated transport system with a heavy-duty trailer link and a traction link corresponding to it in terms of coupling capabilities [9,10]

As an analogue of the load capacity, we take a dump truck BelAZ - 75214 from table 1 and a load coefficient of a heavy trailer $k_{Tr} = 4$ from table 2.

Determination of the total weight of the trailer.

Carrying capacity trailer:

$Q_{Tr} = 1\ 800$ kN.

Own weight trailer:

$$G_{Tr} = \frac{Q_{Tr}}{k_{Tr}} = \frac{1800}{4} = 450 \text{ kN} \quad (3)$$

Total trailer weight:

$$G_{Ttw} = Q_{Tr} + G_{Tr} = 1\ 800 + 450 = 2\ 250 \text{ kN} \quad (4)$$

Determination of the force required to tow a loaded trailer.

Road conditions in open pit mining:

- rolling resistance coefficient, f = 0,02;
- steering bias, i = 7 – 10 % (corresponds to the angle of rise $\alpha \approx 6^0$);
- operating speed of the transport system, v = 10 km/h = 2,78 m/s.

The strength of the total resistance to the movement of the trailer:

$$P_{\psi Ttw} = G_{Ttw}(f \cdot \cos\alpha + \sin\alpha) \quad (5)$$

$$P_{\psi Ttw} = 2250 \cdot (0{,}02 \cdot 0{,}99 = 0{,}174) = 436{,}05 \text{ kN} \quad (6)$$

For uniform rectilinear movement, a force must be applied to the towing device of the trailer (P_{kr}), equal to the strength of the total resistance to movement:

$$P_{kr} = P_{\psi Ttw} \quad (7)$$

Determination of the power spent on towing a trailer:

$$N_{Ttw} = P_{Ttw} \cdot v \quad (8)$$

where N_{Ttw} – trailer towing power, kW.

$$N_{Ttw} = 436{,}05 \cdot 2{,}78 \approx 1212{,}3 \text{ kW} \quad (9)$$

Determination of tractor engine power:

$$N_e = \frac{N_{Ttw}}{\eta_t} \quad (10)$$

where N_e – tractor engine power, kW;

$\eta_t \approx 0{,}8$ – transmission efficiency.

$$N_e = \frac{1212{,}2}{0{,}8} = 1515{,}25 \text{ kW} \quad (11)$$

Taking into account the power reserve for overcoming unforeseen resistances, the effective engine power should be taken 8 - 10% higher than the calculated:

$$N_{ef} = N_e + 10\ \% = 1668 \text{ kW} \quad (12)$$

The calculated power corresponds to the engine power of the dump truck BelAZ - 75214.

Choose the weight of the tractor from the condition of lack of slipping of the drive wheels.

The condition of the movement of the transport system in the absence of slipping of the wheels:

$$P_t \leq P_\varphi \quad (13)$$

where P_t – tangential traction on tractor wheels, kN;

P_φ – traction force of wheels with soil (traction force on coupling), kN.

$$P_\varphi = G_\varphi \cdot \varphi \tag{14}$$

where G_φ – coupling load of the tractor (weight per drive wheel), kN;
φ = 0,6 - 0,7 – coefficient of coupling for open cast mining.

For full drive systems:

$$G_\varphi = G_T \tag{15}$$

Accept the boundary condition for coupling:

$$P_t = P_\varphi \tag{16}$$

To be able to tow a trailer, it is necessary that the tangential traction force P_t be not less than the total force of resistance to the movement of the trailer:

$$P_t = P_{\psi Ttw} = P_{kr} \tag{17}$$

where P_{kr} – traction force on the hook of the tractor.

Then, in view of the above:

$$G_T = \frac{P_{\psi Ttw}}{\varphi} = \frac{436{,}05}{0{,}6} \cong 727 \text{ kN} \tag{18}$$

4. Results of the experiment and their analysis

Compare the existing dump truck with a carrying capacity of 1800 kN with the projected weight indicators:

BelAZ – 75214:
own weight, kN - 1570;
factor carrying capacity - 1,15.
Projected:
trailer weight, kN – 450;
tractor weight, kN – 727;
total weight, kN - 1177;
factor carrying capacity – 3,09.
Weight difference, kN – 393.

That is, the movement of cargo is carried out by a transport system having its own weight 25% less than the base.

5. Conclusion

As indicated earlier, the weight of the tractor is selected for reasons of lack of slipping of the drive wheels. Having reduced the traction force realized on the drive wheels, transferring part of the power to the trailer wheels (i.e., making them drive - AATS), the total traction force required for the movement of the ATS does not change. This allows you to reduce the weight of the tractor by an amount proportional to the power given to the trailer [7,8]

Thus, based on the above calculations, it can be concluded that it is advisable to develop a set of machines for opencast mining on the basis of ATS and AATS.

References

[1] Bazhenov EE, Golomidov SI and Yushkov AI 1985 Determination of the drive power of the active axle of trailers *Exploitation of timber transport rolling stock: interuniversity*. Sat. sci. tr. (Sverdlovsk) p 72-75
[2] Platonov VF 1989 All-wheel drive cars (Moscow: Mechanical Engineering) p 312
[3] Polungyan AA 2009 Design of all-wheel drive vehicles (Moscow: MSTU them. N.E. Bauman) p 488
[4] Bazhenov EE 2009 Articulated transport and technological systems (Ekaterinburg: USTU – UPI) p 174
[5] Bazhenov EE, Buynachev SK and Kruchinin IN 2010 Fundamentals of the theory of articulated transport systems (Yekaterinburg. UrFU) p 257
[6] Bazhenov EE 2010 Modular principle of modeling of articulated transport-technological systems *Tractors and agricultural machinery* (Ekaterinburg) p 20
[7] Bazhenov EE and Vlyukhin AV 2010 Application of articulated transport systems in the extractive industries of the national economy *Truck: Scientific and Technical. magazine* **4** (Moscow) p 34

[8] Bazhenov EE 2010 Modular principle of the synthesis of transport and technological systems *Izvestiya VUZov. The Volga region. Technical sciences* **2** (Penza) p 79
[9] Dhir A and Sankar S 1997Analitical model for dynamic simulation of off-road Tracked Vehicles *Vehicle system dynamics* **27** (USA) p 37
[10] Watanabe K and Kitano M 1986 Study on streerability of articulated tracked vehicles *Journal of Terramechanics* **23** (USA) p 69

IOP Conf. Series: Materials Science and Engineering **695** (2019) 012002 doi:10.1088/1757-899X/695/1/012002

Experimental research of the electromechanical steering system

A V Tumasov, D Yu Tyugin, P O Beresnev, D M Porubov, V I Filatov

Nizhny Novgorod State Technical University n.a. R.E. Alekseev, Minin str., 24, Nizhniy Novgorod, 603950, Russian Federation

E-mail: pavel.beresnev@nntu.ru

Abstract
The development of advanced driver-assistance systems (ADAS) is a global trend. One of the types of ADAS are systems that control the movement of a vehicle in a lane, preventing unintended departure from the lane. When implementing the lane keeping assist system, one of the most important functions is the implementation of external steering control. This article presents tests of the electromechanical steering system with the possibility of external control. Conducted experimental research of the developed system. Based on the test results, conclusions were made on the system's compliance with the requirements of GOST 31507-2012, and recommendations were given on the applicability to vehicles of the GAZ Group, in particular, the lane keeping assist system.

Keywords: ADAS, electromechanical steering system, lane keeping assist system, light commercial vehicles

1. Introduction
The implementation of driver assistance systems (ADAS) helps to improve the road safety of vehicles. The lane keeping system is a type of ADAS that prevents unintentional lane departure. When designing a lane keeping system, one of the most important functions is the implementation of external steering control.

The lane keeping system is planned for installation on vehicles of the GAZ Group, which are equipped with a rack-and-pinion steering gear with hydraulic booster. It was proposed to use a steering mechanism with an electromechanical power steering. The development of this steering mechanism is presented in [1].

To control the electromechanical power steering, algorithms, software, and a control unit have been developed [2]. The algorithm of the control unit allows to implement three operation modes of the steering system:

- electric power steering mode;
- external steering mode;
- emergency mode (manual control mode).

The developed steering mechanism with an electromechanical power steering, the software and the control unit were installed on the GAZelle Next prototype and a number of tests were carried out on the compliance of this steering with the requirements of GOST 31507-2012 standard [3].

2. Description of vehicle's tests
The tests were carried out according to the developed program and test methodology of the prototype with the possibility of external control. The purposes of the tests were: verification of the design and circuit decisions of the developed steering, confirmation of the final decisions made on the basis of assessing the functioning of the steering control on the car. As a testing venue, the GAZ Group's test site was used. The test conditions correspond to those specified in paragraph 5.2 of GOST 31507-2012. The test location is shown in Figure 1.

IOP Conf. Series: Materials Science and Engineering **695** (2019) 012002 doi:10.1088/1757-899X/695/1/012002

Figure 1. Test site

The first parameter to be determined is the steering force. PCAN-USB X6 CAN-FD (CAN Adapter) is used to read data from the vehicle's on-board network [4]. As the measuring equipment, the Racelogic VBOX 3i RTK multifunction speed meter kit [5] and the CORRSYS-DATRON measuring steering wheel with MSW / S sensor and ZUB-MSW adapter [6] were used. The connection diagram of the measuring equipment is shown in Figure 2.

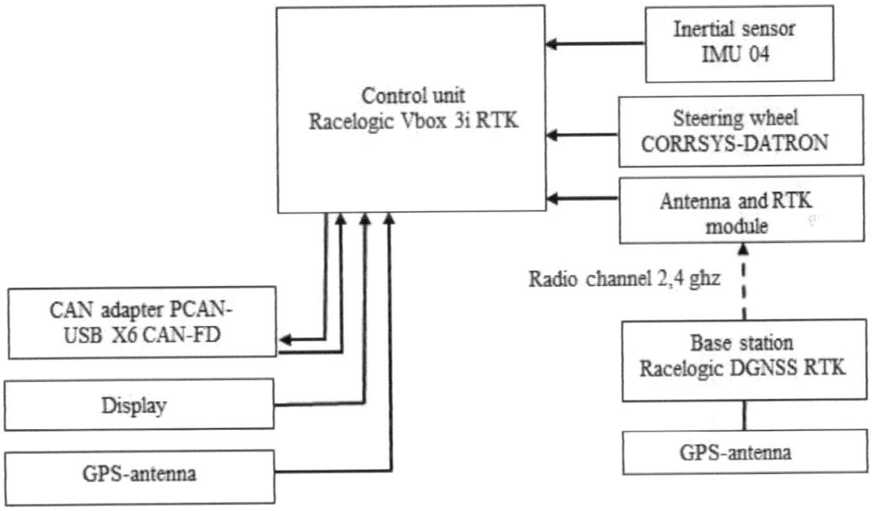

Figure 2. Connection diagram of measuring equipment

During the tests, following parameters were recorded: vehicle location coordinates (speed); steering wheel angle on CAN bus; CORRSYS-DATRON measuring steering wheel angle with MSW / S sensor; torque applied to the measuring steering wheel on the driver's side; steering wheel speed; time. The location of the measuring equipment on the vehicle is shown in Figure 3.

At the first stage, the steering force was measured on a stationary vehicle with the electric power steering running. Before starting the test, the position of the measuring steering wheel was calibrated relative to the position of the steering wheel, during the tests the discrepancy between the values of the sensor for the position of the steering wheel and the sensor for measuring the steering wheel did not exceed 1 °.

IOP Conf. Series: Materials Science and Engineering **695** (2019) 012002 doi:10.1088/1757-899X/695/1/012002

Figure 3. Location of the measuring equipment inside the vehicle

To determine the effort on the steering wheel with a stationary vehicle, the steering wheel was slowly turned from neutral to the extreme right position, and then to the extreme left position, the maximum value of the steering angle was 612 °. From the graph shown in Figure 4 it can be seen that the maximum force exerted on the steering wheel by the driver does not exceed a value of 50 N, based on which it can be concluded that the object passed the test. In this case, the peak forces in the extreme positions of the steering wheel were not taken into account, since the effort on the driver's side increased without turning the steering wheel.

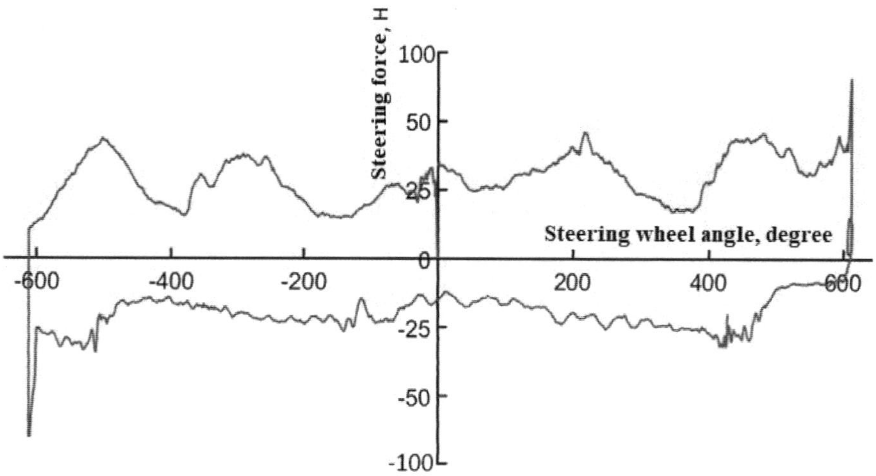

Figure 4. Steering effort graph

2.1. Determination of the effort on the steering wheel (with a working power steering: entrance to a turn with a radius of 12 m at a speed of at least 10 km/h)

The next stage of the tests was the determination of the effort on the steering wheel during the “turn-in” maneuver. On a flat horizontal platform, a circle with a radius of 12 m was marked using road cones (Figure 5).

Figure 5. Circle marked by road cones with a radius of 12 m

During this test, the speed of vehicle's motion clockwise and counterclockwise around a circle with an internal radius of 12 meters was not lower than 10 km / h, which is confirmed by the graphs presented in figures 6 and 7.

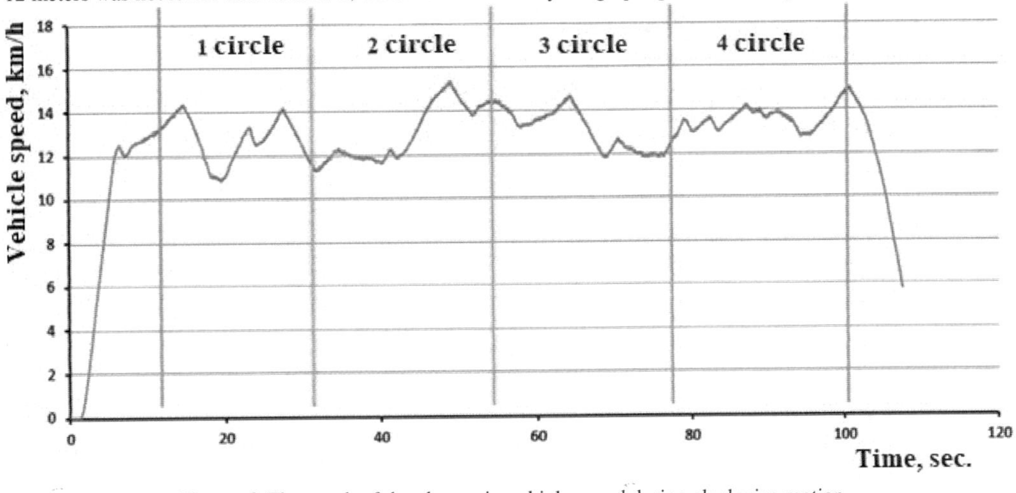

Figure 6. The graph of the change in vehicle speed during clockwise motion

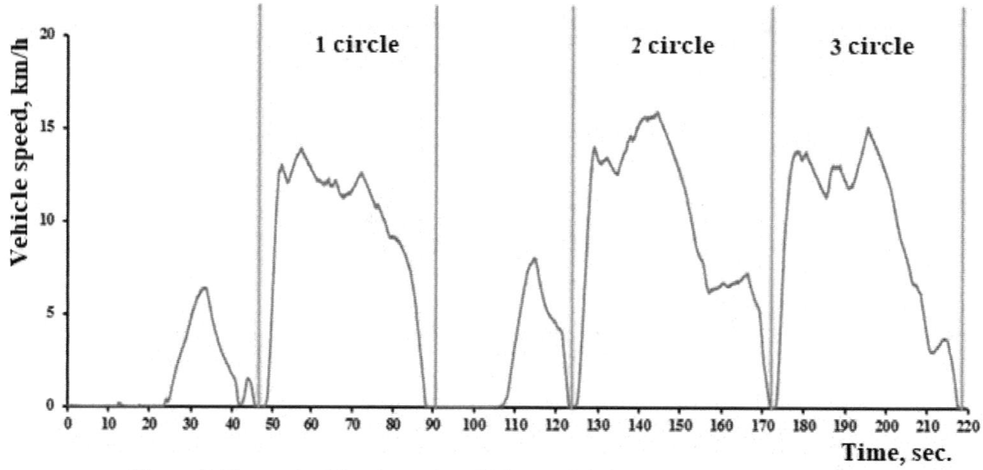

Figure 7. The graph of the change in vehicle speed during counterclockwise motion

To determine the indicators of effort on the steering wheel, the test driver accelerated the car to a speed of at least 10 km/h. At the entrance to the test section, the steering wheel was rotated at a constant angular speed alternately: to one side to the extreme position, and then to the other side also to the extreme position. The value of the steering speed was set in such a way as to ensure the transition of the car within 4 seconds from a rectilinear movement to movement in a circle whose overall radius is 12 meters. According to GOST 31507-2012, the test object is considered to have passed the test if the measured values of the effort on the steering wheel of the vehicle do not exceed the nominal values of 150 N. Figures 8 and 9 show the change in effort on the steering wheel when moving clockwise and counterclockwise, respectively. According to these graphs, we can conclude that the maximum effort on the steering wheel does not exceed 65 N when moving clockwise and 50 N counterclockwise.

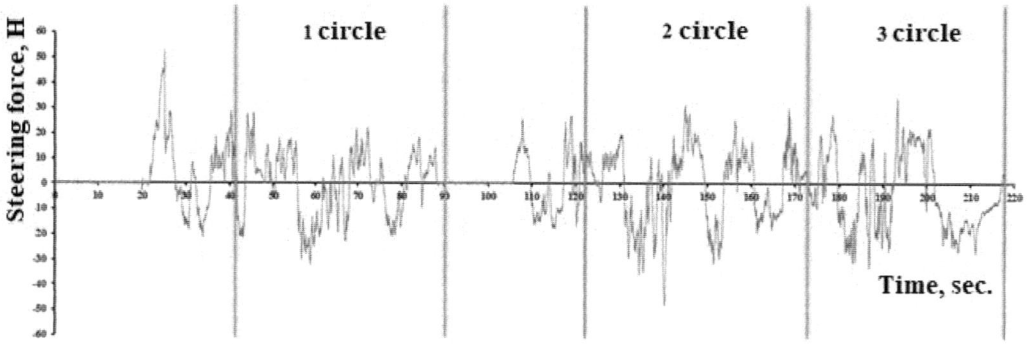

Figure 8. The graph of the change in effort on the steering wheel when moving counterclockwise

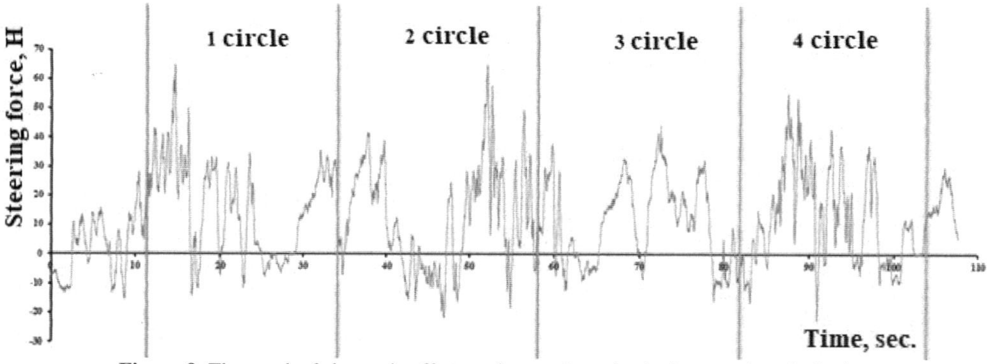

Figure 9. The graph of change in effort on the steering wheel when moving clockwise

2.2. Determination of effort on the steering wheel (inoperative power steering: entry into a turn with a radius of 12 m at a speed of at least 10 km / h)

At the second stage, the same test site was used as at the first stage (the circle with the radius 12 m). During this stage, the vehicle speed was also not lower than 10 km/h. Pictures with test fragments are shown in Figures 10 and 11.

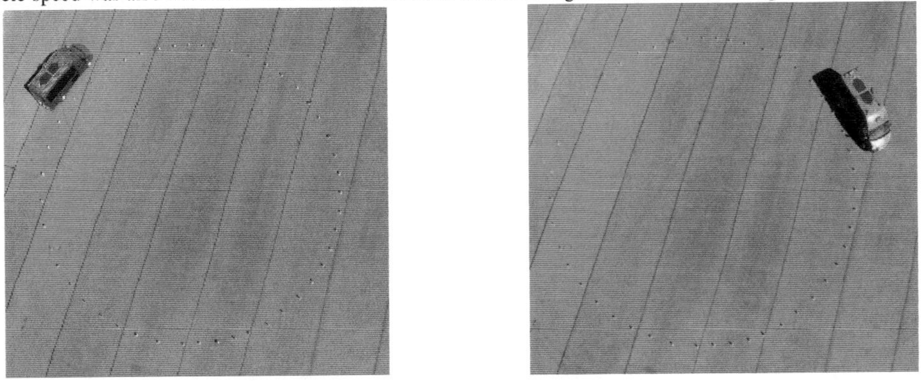

Figure 10. Clockwise motion

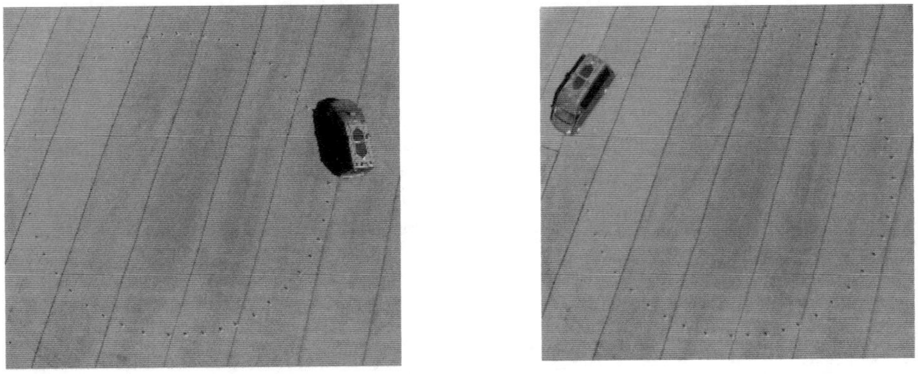

Figure 11. Counterclockwise motion

Indicators of effort on the steering wheel were determined using the methodology described in the previous paragraph, except that there were simulated a malfunction of the power steering by disconnecting the power supply of the electric motor. According to GOST 31507-2012, the force on the steering wheel without the power steering must not exceed 300 N. The graphs of the changes in the force on the steering wheel without power steering are shown in Figures 12 and 13.

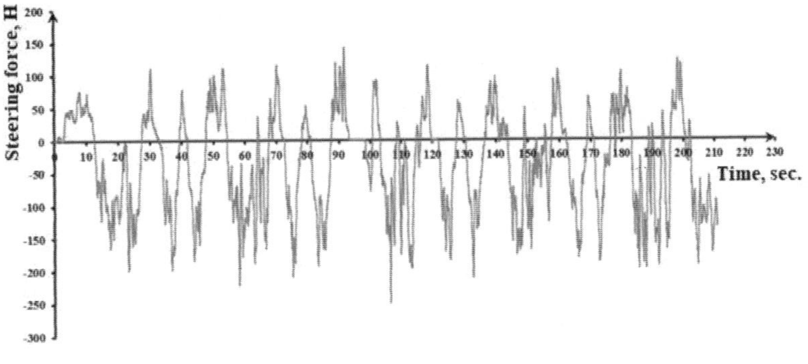

Figure 12. The graph of change in effort on the steering wheel when moving counterclockwise

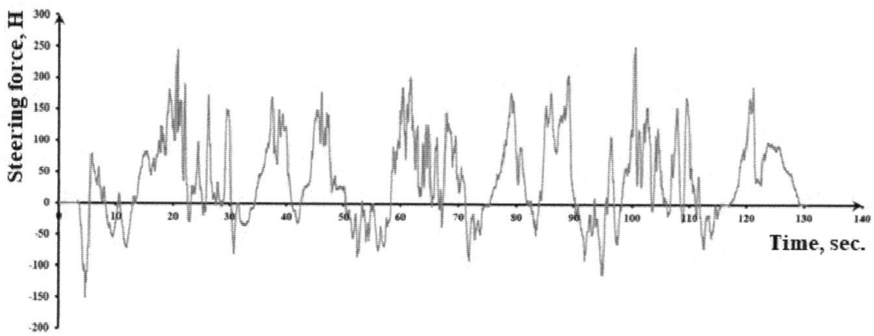

Figure 13. The graph of change in effort on the steering wheel when moving clockwise

From the obtained results, it can be concluded that the force on the steering wheel does not exceed 250 N when moving counterclockwise and clockwise.

3. Steering stabilization when leaving a circle with a radius of 50 m

At the third stage of testing, the vehicle was located on a flat horizontal surface. At this stage, the test driver accelerated the vehicle to a speed of 51 km/h, after which the car was switched from straightforward movement to circular motion, the overall radius of which is 50 meters, with the vehicle leaving the circular path after the steering wheel rim is free of efforts exerted by the test driver. The speed was constant with an error of ± 2 km/h (Figure 15) until the rotation of the steering wheel were stopped within 6 seconds, after which braking was performed until the vehicle stopped completely.

Figure 14. Test fragments when exiting a turn with a radius of 50 meters

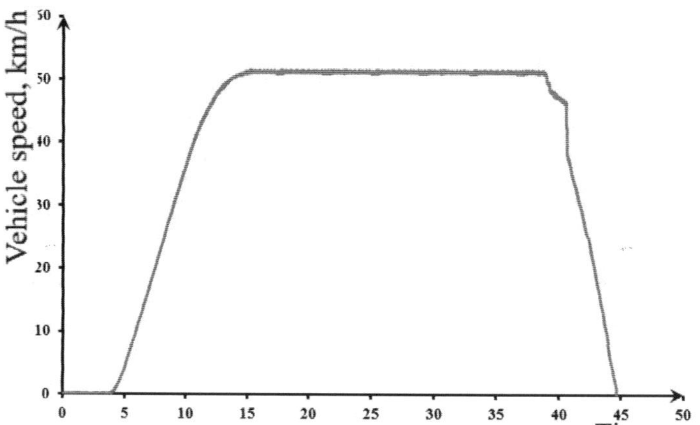

Figure 15. The average value of the speed during the tests

After 6 seconds after the steering wheel was released, its angle from the neutral position was recorded. Three runs were made in each direction with the steering wheel turning clockwise and counterclockwise. Figure 16 shows the average value of the angles of the position of the measuring steering wheel based on the results of all six runs. The position of the steering wheel was determined by the readings that were transmitted to the CAN bus from the steering wheel position sensor (SAS).

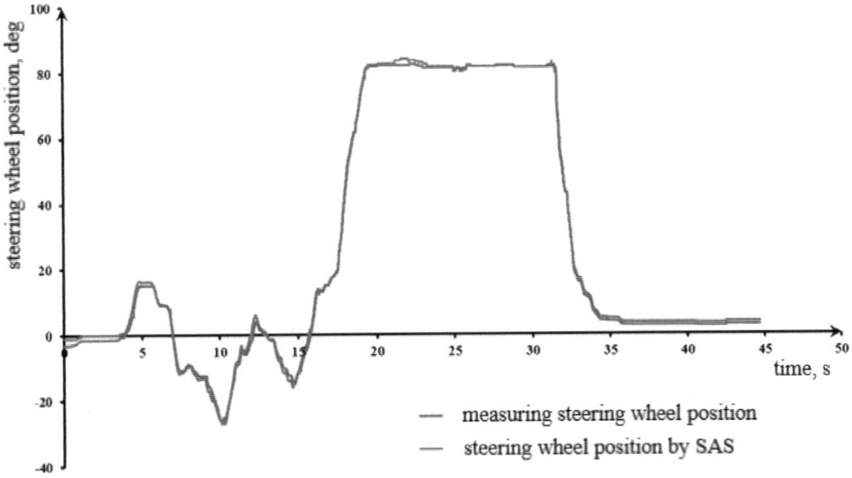

Figure 16. The average angle value of the steering wheel during tests

The graphs show that after the driver has released the steering wheel rim, the steering wheel returns to the neutral position without going through the neutral position and without oscillatory processes. The residual value of the steering wheel position is 6 degrees, which is 7.5% of the position value at the moment the steering wheel is released (80 degrees). The difference between the values of the angles of the position of the measuring wheel and the position of the steering wheel according to the SAS sensor are in the range of 1 degree.

4. Tests of external control mode

During the fourth test stage, to determine the steering angle in the external control mode, the vehicle was located on a flat horizontal asphalt surface. Correspondence of the steering wheel angles of the vehicle to the control signal of the data bus was made by recording the readings of measuring equipment synchronized in time with a log file recorded using a CAN adapter connected to the Racelogic VBOX 3i RTK and the measuring steering wheel CORRSYS-DATRON with MSW / S sensor and ZUB-MSW adapter, according to the diagram in Figure 2. At this stage, the vehicle speed was zero.

The angle of rotation of the steering wheel (wheels of the steered axis) was determined as follows: on a stationary car there was a turn of a steering wheel from a neutral position to the right with angle of 360 degrees, a turn was made by sending CAN messages; then the position of the steering wheel was fixed, then on a fixed car a turn of the steering wheel was made from the right position equal to 360 degrees, to the left position of the steering wheel equal to -360 degrees. To exclude the influence of external factors, after each turn of the steering wheel, the vehicle was moved 0.4-0.6 meters forward. The test was performed three times.

From the graph of the correspondence of the angle of rotation of the steering wheel to the control signal are shown in Figure 17. Tt can be concluded that the deviation during the test of the set and actual value of the angle is not more than 1 degree.

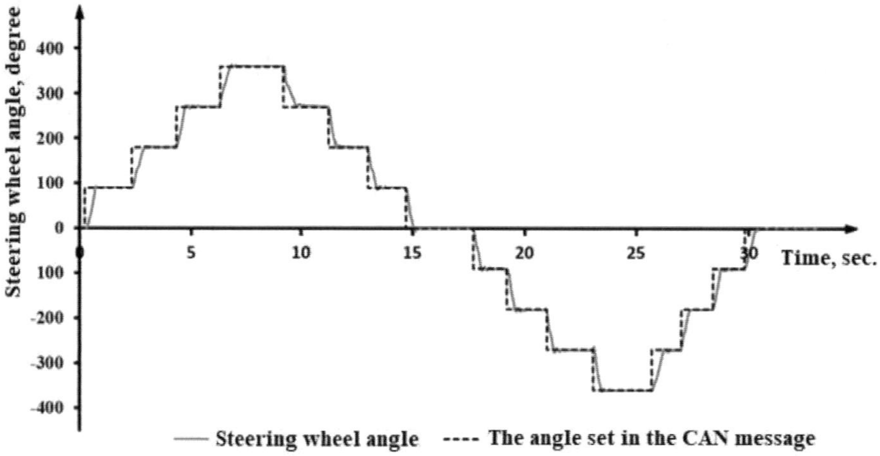

Figure 17. A graph of the correspondence of a given and measured position of the steering wheel

Steering wheel speed in the external control mode (stationary car)
To analyze the time when the power steering reaches the required value of the angle of rotation of the steering wheel, after processing the results obtained earlier, the time interval was fixed from the moment the control signal was sent to the data bus until the steering exited to the specified value of the angle of rotation. As a result of the tests, the average steering speed was 173 deg/s, the maximum turning speed was 245 deg/s. Also, during the processing of test data, it was found that in the steering there is practically no delay in the reaction of the actuator to the control signal via the CAN bus, and is less than 0.005 seconds.

5. Conclusion

Based on the results of these tests, it can be concluded that the developed electromechanical power steering with external control meets the requirements of GOST 31507-2012.

Namely, the forces exerted on the steering wheel in motion do not exceed the maximum permissible value both with a running power steering 65 N, with a permissible 150 N, and in emergency mode without an electric power steering 250 N, with a permissible 300 N. And in a static position the force does not exceed a value of 50 N, with a permissible 60 N.

Also at the exit from a turn with a radius of 50 meters and when the steering wheel rim is released, the steering wheel returns to a neutral position without going through it and oscillatory processes. The residual angle value does not exceed the value of 8% with a permissible 30%.

Regarding external control, it can be concluded that the turning speed of the electric power steering is sufficient for the entire range of operating speeds of the vehicle. A delay in the reaction of the actuator to control commands in 0.005 seconds will allow the system to quickly make adjustments to the motion path if necessary.

Thus, we can conclude that the developed steering is applicable for the implementation of driver assistance systems [7] on GAZ Group vehicles of the NEXT Gazelle family.

Acknowledgements

This research done with the financial support from Ministry of Education and Science of the Russian Federation in the frame of the complex project "The establishment of the high-tech manufacturing of safe and export-oriented GAZ vehicles with autonomous control systems and the possibility of integration with the electric platform on the base of components of Russian production" under the contract №03.G25.31.0270 from 29.05.2017 (Governmental Regulation №218 from 09.04.2010).The experimental research was conducted with the use of measurement equipment of the NNSTU Centre of collective using "Transport Systems".

References

[1] Porubov D.M. Automated vehicle traffic control system based on recognition of the road scene and its objects / Porubov D.M., Beresnev P.O., Tyugin D.Yu., Tumasov A.V., Belyakov V.V., Zezyulin D.V. // Proceedings of the Moscow State Technical University MAMI. - 2018. - No. 1. - S. 52-63

[2] Porubov D.M. Development of a lane keeping system for commercial vehicles / Porubov D.M., Pinchin A.V., Tyugin D.Yu., Tumasov A.V., Beresnev P.O., Belyakov V.V. // Transactions of NSTU named after R.E. Alekseeva. - 2019. - No. 2. - S. 197-204

[3] GOST 31507-2012. Motor vehicles. Manageability and stability. Technical requirements. Test methods // Moscow: Standartinform

[4] PCAN-USB X6 CAN-FD. Retrieved July 18, 2019, from: https://www.peak-system.com/PCAN-USB-X6.438.0.html?&L=1

[5] Racelogic VBOX 3i RTK. Retrieved July 18, 2019, from: https://www.vboxautomotive.co.uk/index.php/en/products/data-loggers/vbox-3i-dual-antenna-rtk

[6] Datron Technology. Retrieved July 18, 2019, from: https://www.datrontechnology.co.uk/ (accessed: 07/18/2019).

[7] Porubov, D. M., Groshev, A. M., Mishustov, V. P., Palutin, Y. I., & Tumasov, A. V. (2018). Development of technical solutions for development of a commercial vehicle driver assistance systems. Paper presented at the IOP Conference Series: Materials Science and Engineering, , 386(1) doi:10.1088/1757-899X/386/1/012011

Researching rolling resistance of the wheeled forestry tractor at skidding

Vladislav Klubnichkin[1], Evgeny Klubnichkin[1], Vladimir Makarov[2] and Maxim Podrubalov[1]

[1] Department of transport and technological means and equipment of the forest complex, Mytishi branch. Bauman Moscow State Technical University, Russian Federation, Mytishi
[2] Nizhny Novgorod State Technical University n.a. R.E. Alekseev, Minin str., 24, 603950, Nizhny Novgorod, Russian Federation

E-mail: vklubnichkin@gmail.com

Abstract
Off-road performance improvement remains one of the main problems in designing wheeled dragging and forestry tractors. Improvement of off-road performance of the dragging and forestry tractors is closely related to the general problem of wheeled vehicles' off-road performance, however it features a number of specific aspects which require specialized research to be made.

During tree skidding by a wheeled tractor, vertical load on a double-reduction axle greatly changes that is linked to additional load by a roundwood bundle and redistributed in-motion load. The above in-motion redistribution of load and driving moments along the axles greatly affects the vehicle's off-road performance. While skidding, the increased load on a wheel leads to increased specific pressure, track depth and rolling resistance.

This work employs simplified tire calculation model used as a base to determine relationship between load and deformation. Calculation is made successively for each elementary tire section that allows for finding diagram of pressure distribution in the contact longitudinal plane and researching rolling resistance related to rutting and hysteresis losses in the tire material. Further experimental research of the driving wheel and wheel system motion allowed for elaborated theoretical analysis and helped to justify parameters the forestry tractor wheels.

The following tasks were set during this work:

1. To make efforts to elaborate approximated method for estimating relative elongation of the rolling resistance depending on wheel parameters, vertical load and air pressure. The method is based on calculation of contact pressure distribution on the basis of simplified tire model
2. To conduct experimental research of rolling resistance for one wheel and wheel set of a tractor in the leading movement mode with due regard to tire tangential deformation.
3. To study aspects of a process of rolling resistance of a forestry tractor during tree skidding in summer time. To run comparative draw tests of a forestry tractor equipped with tires of variable size.
4. To find out possibilities to reduce rolling resistance basing on theoretical and experimental researches.

Keywords: Wheeled forestry tractor, Rolling resistance, Tree skidding, Off-road performance

1.Introduction

Today forest industry and forest management of the Russian Federation have a mission of sustainable use of the national forest wealth. This mission can be carried out through designing a set of equipment for gradual and selective cutting, clean cutting, regeneration and forestry works, as well as fire-fighting and pest management(sandu), .

Those types of cutting feature low volume of cut wood per unit of area. It varies from 7 m^3 per 1 ha at thinning from 50 to 90 m^3 during gradual and selective cutting. The equipment used during such cutting and forestry works shall meet such important requirements as high mobility and less damage to undergrowth and left trees. In this condition use of wheeled tractors is more reasonable as compared to tracked vehicles.

Off-road performance improvement remains one of the main problems in designing wheeled dragging and forestry tractors. Improvement of off-road performance of the dragging and forestry tractors is closely related to the general problem of wheeled vehicles' off-road performance, however it features a number of specific aspects which require specialized research to be made.

Performance research needs solving of a number of issues and involves examination of rolling resistance, wheel friction features and soil bearing capacity.

The most prominent researches in this filed are made by Ya.S. Ageykin, V.F. Babkov, M.G. Bekker., A.K. Birulya, N.F. Bocharov, V.A. Deligovskiy, V.P. Goryachkin, , V.L. Knoroza, V.M. Kotikov, A.F. Poletaev, Yu.E. Sharikyan.

A dragging tractor during ground skidding has a higher trailing weight, with that the load on the wheels greatly changes both due to bundle weight and re-distribution of weight between the vehicle axles, especially during the movement (Melekhov, 1980; Pobedinsky and Zheldak, 1989; Kotikov, 1995). During off-road movement in the forest such change in wheel loads may lead to tire overload, sharp rise in ground pressure and increased rolling resistance. Rolling resistance frequently constitutes the majority of the total movement resistance and requires priority research.

Researching impact of changes in wheel vertical load on rolling resistance demanded review of load transfer from a wheel rim to its base, its deformation and pressure distribution in the contact area. Strictly theoretical solution to this issue meets huge difficulties, and when wheel rolling mechanics are examined, use is made of approximate methods based on calculation of tire simplified models. This work reviews a model pneumatic tire which allowed finding a dependence between rolling resistance and forestry vehicle wheel parameters, tire air pressure and vertical load.

Rolling resistance research is inextricable linked with research of operation of the wheel system – in movement distribution of load between axles and related re-distribution of torques, as well as finding of optimum weight distribution of the dragging tractor. Such researches are needed for solving a whole range of practical tasks related to designing of cross-country wheeled running gear and selection of optimum design of the vehicle in general (Bekker, 1969; Wong, 2009, Horn 2007).

2. Researching a process of rolling resistance of a forest vehicle wheel with pneumatic tire

Air pressure forces and tire external load may be simply taken up by the cord threads located in one radial plane. We call this form of carcass radial-balanced. Material reactions in the planes different from a radial one can take up air pressure in case of disordered radial-balanced form of the shell, which is used to define them and consider in the calculations. Proceeding from this provision, the tire calculated model represents a carcass where the cord threads are located strictly in radial plane, with that the model carcass dimensions are reduced from radial-balanced form to the dimensions of similar to the calculated tire through the action of a breaker strip (Fig. 1) (Kotiev et al. 2009; Pirnazarov et al. 2015).

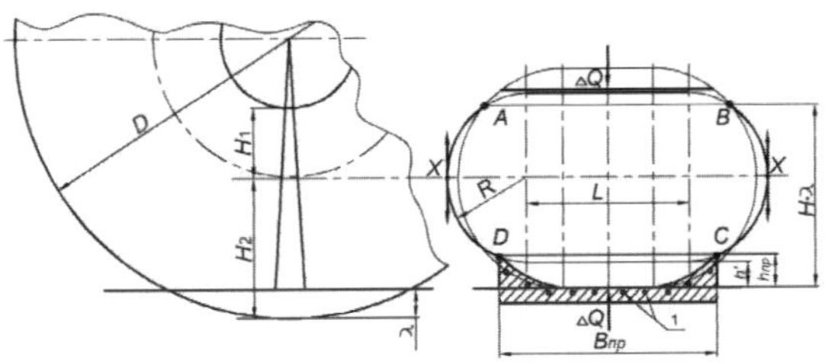

a) b)

1 – Breaker strip

Figure 1. Calculation diagram

The given in this work solution of differential equations of flexible thread tension by action of air pressure and widely known experimental data determine that the flexible thread is deformed over constant radius (Goberman et al. 2003; Kochnev and Bocharov et al. 1974). Let us review an equilibrium of forces and reactions in an elementary section. A sum of vertical components of forces of air pressure and external load on the X-X section is equal to $P_0 \cdot (B + {}_\Delta B) - {}_\Delta Q$. Vertical reactions of the flexible thread tension along X-X axis are equal to product $2R \cdot P_0$. For the given section the width in length units is not considered in the calculation. The following dependence is obtained from the condition of equilibrium of forces and reactions:

$$P_0(B + {}_\Delta B) - {}_\Delta Q = 2R \cdot P_0,$$

$${}_\Delta Q = P_0 \cdot (B + {}_\Delta B - 2R), \tag{1}$$

Calculation of loading consists in calculation of a tire side curvature radius R and increase of width ${}_\Delta B$. We initially assume that the elementary section (Fig. 2 a) is cylindrical (hose section), rimless, absolutely flexible, inextensible, the perimeter width is even and equal to the length unit (1 cm), the tangential forces – reactions of carcass tension, perpendicular to radial, are absent. For our purpose we call the section a separate flexible thread. External load is transferred to flexible thread via two horizontal planes M, N.

Since load is transferred via parallel planes and the thread is deformed along the radius, the side form represents a semicircle.

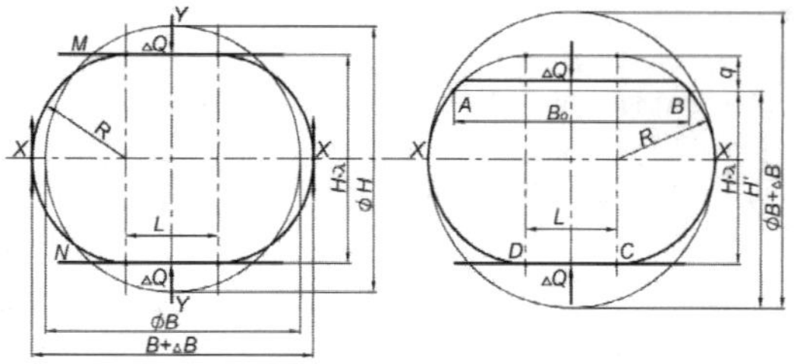

a) b)
Figure 2. Tire elementary radial section

Ideally an elementary section has a form of circle (hose section), is absolutely flexible, inextensible, tangential forces are absent. For our purpose we call a section a separate flexible thread. External load is transferred to a flexible thread via two parallel planes. For the assumed conditions $2R = H - \lambda$, and the section width increase under load and length of thread contact length L are found on the assumption that the thread is inextensible and its height is equal to width $H=B$.

Relationship between the load and deformation becomes as follows

$$\Delta Q = \pi/2 \cdot P_0 \cdot \lambda \tag{2}$$

According to equation (2) relationship between the load and deformation is straight-line and may be expressed through specific rigidity. When the load is taken by tire air pressure only (carcass effect is not considered), the load taken by the section can be found by the volume of compressed air, passing through calculation of area and form of contact surface or base rigidity. Then pressure distribution is determined by section areas in contact.

Hence, specific pressure shall be equal to air pressure. However, specific pressure is equal to air pressure only when the section between parallel planes is deformed. In this case equation (2) becomes as follows

$$\Delta Q = P_0 \cdot L. \tag{3}$$

If the contact surface is curvilinear (Fig. 3), then specific pressure is not equal to air pressure. Such deformation of a shell occurs under influence of the tire tread or when a section is deformed on the crushable soil. Definition of relationship between load and radial deformation reduces to calculation of a radius of the side curvature and impact of the carcass on the tire specific rigidity that allows for finding distribution of specific pressure both on the hard foundation and crushable soil.

Through sequential consideration of impact of a rim, tread on the tire side curvature radius, as well as impact of tangential forces and carcass rigidity we find relationship between load on radial section and deformation, which is used to calculate specific rigidity. In case of hard foundation, the value of deformation of each wheel section in contact is easily calculated, hence pressure profile along contact length can be found.

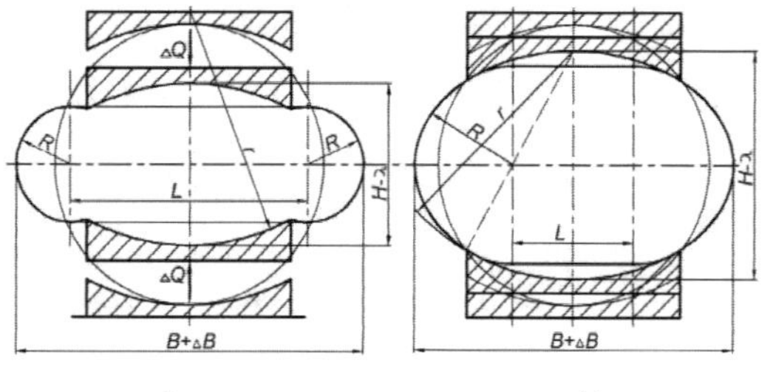

a) b)
Figure 3. Calculation diagram to find tire deformation

Elementary section with deformation $2/3\lambda$ features average lifting capacity along the contact length. Hence load Q taken by the entire tire at deformation and specific rigidity γ are equal:

$$Q = l \cdot \Delta Q_{av} = 2\sqrt{D\lambda} \cdot \Delta Q_{av}; \tag{4}$$

$$\gamma = \frac{3}{2} \cdot \frac{Q}{2\sqrt{D\lambda} \cdot \lambda},$$

Analysis of the above relationships demonstrates that with equal deformation of the tire radial element sides on the solid foundation and crushable soil the vertical loads on this element slightly differ, although the contact areas and specific pressures are essentially different. Notwithstanding that the tire specific rigidity cannot be considered a constant value, the term can be applied to calculate tire specific pressure on the soil in the method of transfer from specific pressure on solid foundation to specific pressure on soil by knowing the difference between the areas of contact at equal deformation

IOP Conf. Series: Materials Science and Engineering **695** (2019) 012003 doi:10.1088/1757-899X/695/1/012003

of sides of the tire radial element.

Thus, even if the rigidity of the entire tire depends on rigidity of the foundation, it can be recalculated to specific rigidity which within the certain limits can be taken as a tire parameter not dependent on foundation rigidity in calculating on-soil rolling.

3. Analytical calculation of rolling resistance on solid crushable foundation

The force of rolling resistance on solid foundation can be found by energy loss in the course of movement. The rolling load is taken by a carcass and air. Let us accept that energy losses for air volume compression can be omitted. In case uniform rolling the tire air volume does not change and its movement is insignificant. Contact friction shall be considered by coefficient $K \approx 1.1$ of the main energy losses for hysteresis. Hysteresis energy losses shall be found by energy taken by the tire carcass in the course of rolling and a hysteresis loss coefficient.

The found relationship between the load taken by the carcass, its deformation, deformation of the cord thread tread is used to find the energy taken by carcass, the losses and then the rolling resistance

$$f_t = \frac{3 \cdot \eta \cdot k \cdot 2\sqrt{D\lambda}}{8D} \cdot \begin{bmatrix} \frac{8}{3} \cdot \left(\frac{E_p \cdot h}{2 \cdot n \cdot i \cdot E_k \cdot cos\beta} + \frac{E_p \cdot \varepsilon \cdot h \cdot 2\sqrt{D\lambda}}{Q} \right) + \frac{2}{3} \cdot \frac{\lambda \cdot E_k \cdot n \cdot i \cdot h^2 \cdot 2\sqrt{D\lambda}}{3 \cdot 0{,}149 \cdot \boldsymbol{B}_1^2 \cdot H \cdot Q} + \\ + \frac{3}{2} \cdot \frac{Q}{2\sqrt{D\lambda}} * \frac{H}{\lambda \cdot n \cdot i \cdot E_k \cdot cos\beta} + + \frac{3}{2} \cdot \frac{Q}{2\sqrt{D\lambda}} \cdot \frac{\psi_1 h_n}{\lambda \cdot B_{tr} \cdot \psi_2 \cdot \psi_{tr} \cdot E_{ph}} \end{bmatrix}. \quad (5)$$

The rolling resistance of the wheel with pneumatic tire rises with increasing hysteresis loss coefficient η, carcass rubber elastic coefficient E_p, carcass width h, tread width h_n. Increased rigidity of the cord layers $n \cdot i \cdot E_k \cdot cos\beta$ leads to reduced loss in the carcass rubber, cord threads, but increased bending loss of the tire sides. The rolling resistance reduces with increased: tire diameter D; tire width B; tread width B_{tr}; elastic coefficient of the tread rubber E_{ph}; as well as tread design ratio ψ_{tr}. The wheel load and air pressure also cause change in the rolling resistance.

The research of tire deformation impact on rolling resistance of the wheel with pneumatic tire on the soil is performed comparison method by moving an equal in size rigid wheel and applying coefficient ψ.

Coefficient ψ shows what fold load Q on the wheel with tire is higher than load Q_{rig} on the rigid wheel at rolling with equal in depth and shape gauge.

$$\psi = \frac{Q}{Q_{rig}} \quad (6)$$

In experimental research the rolling characteristics of the rigid wheel can be obtained during tests with high tire air pressure when its deformation is insufficient and it can be considered by interpolating the research data. A wheel with a rigid rim is considered as a special stamp which defines the soil capacity to bear vertical loads for the given wheel parameters.

Let us successively review interaction between the wheel elements and soil during rolling flowed by rutting. In this case it is possible to define two specific zones for the contact longitudinal plane: front curvilinear and rear planes. The load in the front area successively borne by the wheel sections is increased and depends on the depth of their penetration into soil. The load in the rear plane successively borne by the sections is reduced and depends on the tire parameters and its capacity to distribute contact load.

Basing on the found tire specific rigidity we calculate a relative value - a coefficient ψ which in equal soil conditions describes tire impact on on-soil operation as compared to a similar rigid wheel.

$$\psi = \sqrt{\frac{4b \cdot \forall \cdot E\sqrt{D \cdot \lambda_t}}{3Q_t}} + \sqrt{1 + \frac{4b \cdot \forall \cdot E\sqrt{D \cdot \lambda_t}}{3Q_t}} \quad (7)$$

Relationships for the wheel with pneumatic tire are found with use of coefficient ψ and presented in Table 1.

Table 1 – calculated value of the rolling resistance

Description	Wheel with rigid rim	Wheel with pneumatic tire
Wheel load, Q	$\frac{2}{3} b \cdot c \cdot h\sqrt{Dh}$	$\frac{2}{3} b \cdot c \cdot h\sqrt{Dh} \cdot \psi$
Gauge depth, h	$\sqrt[3]{\left(\frac{1.5 \cdot Q}{b \cdot c \cdot \sqrt{D}}\right)^2}$	$\sqrt[3]{\left(\frac{1.5 \cdot Q}{b \cdot c \cdot \sqrt{D}}\right)^2} \cdot \sqrt[3]{\frac{1}{\psi^2}}$
Rolling resistance force, P_f	$0.86 \cdot Q \cdot \sqrt[3]{\frac{Q}{b \cdot c \cdot D^2}}$	$0.86 \cdot Q \cdot \sqrt[3]{\frac{Q}{b \cdot c \cdot D^2}} \cdot \frac{1}{\psi \cdot \sqrt[3]{\psi}} + P_{ft}$
Rolling resistance, f	$0.86 \cdot \sqrt[3]{\frac{Q}{b \cdot c \cdot D^2}}$	$0.86 \cdot \sqrt[3]{\frac{Q}{b \cdot c \cdot D^2}} \cdot \frac{1}{\psi \cdot \sqrt[3]{\psi}} + f_t$

Coefficient ψ considers impact of the tire deformation on rutting, internal, hysteresis losses are considered by value f_t and P_{ft} defined during rolling with the same load on a hard surface.

Wheel load corresponding to a specific rolling resistance is found by the following formula

$$Q = 1.56(f - f_t)^3 \cdot b \cdot \forall \cdot E \cdot D^2 \cdot \psi^4 \tag{8}$$

Let us convert this relationship by taking all the members which characterize the wheel parameters to the left part, we find

$$\frac{Q}{b \cdot D^2 \cdot \psi^4 \cdot ב} = 1{,}56(f - f_t)^3 \cdot E \tag{9}$$

Analysis of this formula allows for an important conclusion that by observing constant value of relation of the following parameters $\frac{Q}{b \cdot D^2 \cdot \psi^4 \cdot \forall}$ and rolling resistance on a solid foundation $\boldsymbol{f_t}$, we can select wheels of different design which in similar soil environment will feature almost similar rolling resistance. By comparing this relation with its value of the forestry vehicles which meet the specified operation requirements we can evaluate rolling resistance of different forestry vehicles when analyzing their cross-country performance and capability to operate on soil.

4. Experimental research

Unlike agricultural tractors a skidder performs its transportation operation by drawing and by transporting part of load inside a vehicle. A rolling resistance force of the loaded vehicle is frequently more than drawing force used for skidding. Non-uniform distribution of load and torque which occurs during movement with load leads to increase of specific pressure on the track bedding and increase of the rolling resistance. With that the rolling resistance force of the loaded vehicle in the off-road forest environment can be several times more than its value of the empty vehicle.

It is necessary to evaluate parameters of the skidders and equipment considering full efforts and energy costs for load transportation utilized both at rolling and drawing. The relation between full efforts or energy costs for load transportation to general cost of movement is estimated by efficiency of the forest tractor. It can be used to analyze and evaluate operation of a forestry vehicle in view of providing minimum efforts and energy costs for transportation of one cubic meter of timber.

Tractors efficiency η_T is a relation between power used for effective output for load transportation and the engine power:

$$\eta_T = \frac{N_{gr}}{N_e \cdot e} = \eta_m \cdot \eta_\delta - \frac{Q_T \cdot f_{gr} \cdot V}{270 \cdot N_e \cdot e} \tag{10}$$

The power used for transportation of onboard load at rolling is taken proportional to the load weight and forest tractor's weight. The efficiency of the forest tractor takes the following form:

$$\eta_T = \eta_M \cdot (1 - \delta) \cdot (1 - \frac{f_{gr}}{\varphi_0} \cdot \frac{Q_T}{Q_T + \theta_x \cdot K}) \tag{11}$$

Efficiency is a complex indicator which characterizes coupling parameters, rolling resistance and impact on operation mode of the forestry vehicle on the transportation capacity.

To consider full costs both for rolling and drawing of tree bundle, the tractor's towing performance is based not depending on drawbar force bot on wheel full circumferential force. The main evaluation values are respectively: relationship between coefficient of use of hitch weight and skidding, traction coefficient, coefficient of used traction, coefficient of specific trust, as well as rolling resistance for inhaul and outhaul. The above specific coefficients are calculated with due regard to increase of forest tractor hitch weight by tree bundle.

Experimental data allow tracking impact of the wheel load re-distribution at skidding on the rolling resistance.

In the test conditions the tractor's front axle in-motion load was reduced from 2,320 kg to 1,580 kg when skidding a 1.5-m^3 bundle. The rear axle load was increased from 1,580 to 3,140 kg. Front axle unloading coefficient Jj = 0.64, rear axle loading coefficient Jj = 2.11. Weight distribution of the forest tractor from 60% at outhaul has changed to 32% at inhaul. Rolling resistance increased by 1.84 times as compared to outhaul. Tests over crushable soil have uncovered that use of tires 15-24 significantly reduces rolling resistance and improves drawbar-pull properties. The efficiency of the tractor's running gear has improved in the experimental environment from 44 to 55% that is an increase of 11% as compared to operation with use of tires 15-20.

Thus, the rolling resistance of the wheeled skidder during its off-road movement with re-distribution of weight and torque between the axles is largely dependent on the forest tractor load and its distribution between the wheels. Its inhaul values significantly differ from the outhaul or slave movement values and they shall be reviewed separately in the analysis of the forest tractor operation.

Research of the skidder running resistance shall include measurement of the wheel torque and bundle drawing force. This allows approaching to evaluation of improvement of parameters of the tractor running gear and skidding equipment against minimum energy cost for load transportation via efficiency of the tractor and its components. It is necessary to separate forces and required power needed for load transportation by skidding and onboard a tractor as well as of the tractor itself from the total movement costs.

5. Conclusion

Rolling resistance at timber skidding with butt end first constitutes (for off-road capacity limited conditions) the major part of the total circumferential load of the vehicle's wheels.

During timber skidding the rolling resistance depends on the amount of tree bundle that is linked with rise of vertical load and torque on the forest tractor wheels, as well as changed distribution of load and torque between the wheels at inhaul movement.

Tire specific rigidity marginally depends on the foundation rigidity and load and is found theoretically and experimentally by the tire contraction properties on a solid foundation.

Basing on the found tire specific rigidity it has been determined what fold the taken load of the wheel with pneumatic tire is higher as compared to the similar rigid wheel under the same wheel load on the track surface. This value is expressed by coefficient ψ. Value of coefficient ψ for the same soil environment can serve an evaluation characteristic of tire capability to run over crushable soil.

We have derived equations to find relationship between rolling resistance and wheel parameters, air pressure and load (Table 1), applied torque both for a single wheel and wheel system.

It is recommended to justify parameters of the skidder running gear basing on the on-soil tire capacity. On-soil capacity is determined basing on relationship between load and rolling resistance and provides for its necessary value for the given type of forestry vehicles.

Experimental research has found out that optimum weight distribution in the experimental environment was correspondent to the 0.43-0.45 load on the front axle of the tractor's full weight, the rest weight was on the rear axle. Skidding of the front and rear wheels corresponding to the wheel maximum efficiency is different: front wheel skidding is 2-3% more than of the rear wheels which follow the front wheel track.

Increase in load on the skidder wheels by the weight of processing equipment and drawn tree bundle which is 60-80% weight of the running gear as well as specific nature of applied load by the drawn bundle to the tractor pose requirement for a special design of its running gear.

Parameters of the skidder running gear shall be evaluated with due regard for provision of minimum effort and energy costs for skidding.

Experimental tests of the forestry tractor in summer environment on sandy-loam cutting area have found out that during skidding of a 1.5 m^3 bundle the front axle unloading coefficient stood at 0.64, rear axle load coefficient stood at 2.11, in-motion weight distribution on the front axle was 32%, the rest weight was on the rear axle. Under this condition the rolling resistance has increased 1.84 times as compared to the outhaul movement.

Nomenclature

B	width of elementary section without load along the carcass internal surface	[mm]
ΔB	increase in the elementary section width under load	[mm]
P_0	tire air pressure	[bar]
R	radius of elementary section side curvature in points along axis X-X over carcass internal surface	[mm]
ΔQ	external load on elementary section	[N·m]
ΔQ_{av}	lifting capacity of elementary section with deformation $\frac{2}{3}\lambda$	[N]
$l \cong 2\sqrt{D\lambda}$	length of tire contact with foundation	[mm]
γ	tire specific rigidity	-
Q, λ	load and respective tire deformation	[N]
η	hysteresis loss coefficient	-
E_p	carcass rubber elastic coefficient	-
h	carcass width	[mm]
h_n	tread width	[mm]
D	tire diameter	[mm]
B_{tr}	tread width	[mm]
E_{pu}	elastic coefficient of the tread rubber	-
ψ_{tr}	tread design ratio	-
Q_{rig}	load of the wheel with tire and equal rigid wheel at rolling with similar gauge	[N]
ψ	coefficient considering impact of tire deformation on load increase at equal specific wheel pressure on the foundation	-
b	width of wheel in the gauge	[mm]
E	soil deformation module	[MPa]
$\forall$	coefficient which defines soil characteristics for the given wheel parameters	-

Acknowledgements
Research financed by the Ministry of Science and Higher Education of the Russian Federation project number № 9.8996.2017/8.9.

References
[1] Ageykin Ya.S., 1972. Cross-country wheel and combined propulsion.Moscow. Mashinostroenie, Pages 184.
[2] Bekker, M.G., 1969. Introduction to Terrain-Vehicle Systems. The University of Michigan Press, Ann Arbor.
[3] Bocharov, N.F., Tsitovich, I.S., Polungyan, A.A., Semenov, V.M., Tsybin, V.S., Zheglov, L.F., 1983. Designing and calculating high wheeled vehicles patency. Moscow. Mashinostroenie, Pages 299.
[4] Bocharov, N.F., Gusev, V.I., Semenov, V.M., Soloviev, V.I., Filyushkin, A.V., 1974. Vehicles on highly elastic propulsion. Moscow. Mashinostroenie, Pages 208.
[5] Bottinelli, N., Hallaire, V., Goutal, N., Bonnaud, P., Ranger, J., 2014. Impact of heavy traffic on soil macroporosity of two silty forest soils: Initial effect and short-term recovery. Geoderma. Vol. 217–218, 10-17
[6] Cambi, M., Certini G., Neri F., Marchi, E., 2015. The impact of heavy traffic on forest soils. Forest Ecology and Management. Vol. 338, 124-138.
[7] Goberman, L.A., Goberman V.A., Ksenevich, I.P., 2003. Ground trailer transport systems. Technical and economic bases of designing machines and processes. Methodological aspects of project activities and decision making. Volume 3. Encyclopedia. Engineering, Moscow. Pages 788.
[8] Horn, R., Vossbrink, J., Peth, S., Becker, S., 2007. Impact of modern forest vehicles on soil physical properties. Forest Ecology and Management. Vol. 248, Issues 1–2, 56-63.
[9] Kochnev A.M., Anisimov, G.M., 2008. Tests of harvesting machines. SPb.: SPbGLTA, Pages 488.
[10] Kotiev, G.O., Zorin, D.V., 2008. Forecasting the longevity of parts of transport vehicles. The world of transport. Vol. 6. Issue 1. 4-9.
[11] Kotikov, V.M.,1995. Impact harvesting machines on forest soils. Dissertation of the doctor of technical sciences. Moscow state forest university, Pages 214.
[12] Li, L., Sandu, C., 2007. On the impact of cargo weight, vehicle parameters, and terrain characteristics on the prediction of traction for off-road vehicles. Journal of Terramechanics. Vol. 44, Issue 3. 221-238.
[13] Melekhov, I.S., 1980. Forest science. Forest industry. Moscow. Pages 406.
[14] Pirnazarov, A., Sellgren, U., 2015. Reduced testing and modelling of the bearing capacity of rooted soil for wheeled forestry machines. Journal of Terramechanics. Vol. 60, 23-31.
[15] Pobedinsky, A.V., Zheldak, V.I., 1989. Features of thinning in forests with limited forest management. Forestry. Vol. 9. 24-27.
[16] Silaev, A.A., 1972. Spectral theory of suspension of transport vehicles. Mechanical Engineering, Moscow. Pages 192.
[17] Wong, J.Y., 2009. Terramechanics and Off-Road Vehicle Engineering, 2nd Edition. Pages 488.

Researching impact of pressure re-distribution along the length of the track ground contacting area on resistance of the timber-hauling vehicles' travel over snow

Evgeny Klubnichkin, Vladislav Klubnichkin, Alexander Erkhov and Alexey Alyabiev

Department of transport and technological means and equipment of the forest complex, Mytishi branch. Bauman Moscow State Technical University, Russian Federation, Mytishi

E-mail: klubnichkin@mgul.ac.ru

Abstract
The tracked timber-hauling vehicles which operate in the cutting areas having highly deforming soil, virgin snow in particular, feature significantly reduced performance due to rutting and caused frontal resistance.
Most logging enterprises harvest bulk timber in I and IV quarters of the year. However, deep snow during the second half of winter significantly reduces performance of the dragging tractors. Snow mostly affects performance of the feller-skidders and feller-bunchers which need to move from tree to tree to collect load while frequently crossing deep virgin snow.
Snow 0.4-0.6 m deep slightly resists to movement. However, when virgin snow is 0.7 m and deeper, the resistance rises sharply.
The tracked vehicle capacity to cross virgin snow can be improved in two ways: by improving friction properties and by reducing motion resistance. As is evident, more even distribution of track pressure and lack of pressure concentration under individual rollers will improve vehicle's hauling power by reducing frontal resistance.
But as researcher M.G. Bekker stated, despite the fact that pressure distribution was considered one of the most important factors determining vehicle's performance, the theoretical research and assessment of the problem was frequently neglected.
Today designers while designing new tracked dragging vehicles and assessing their performance refer to the researches made for industrial and agricultural tractors. This can be explained by few works dedicated to researching performance of the forest tracked vehicles. The design and operating conditions of the forest and non-forest vehicles are much different so that the results of research made for one type of vehicles cannot be applied to other types of vehicles.
That is why it is necessary to research the process of tracks' impact on snow and pressure re-distribution due to snow reaction to justify parameters of the tracked undercarriage of the forest vehicles, provide optimized location of the gravity center and reduce motion resistance.
Our work is aimed to develop a method to estimate pressure distribution along the length of the track ground contacting area, test the method with experiment and provide recommendations to designers on improving oversnow performance of the forest tracked vehicles.
Keywords: Forest tracked vehicle, Snow, Pressure distribution, Motion resistance

1. Introduction

Interaction between the tracked running gear and crushable soil have been studied by (Anisimov 1975; Bekker 1969; Bodin 1999; Cambi et al. 2015; Dhir and Sankar 1994; Dmitriev et al. 1993; Goberman et al. 2003; Horn et al. 2007; Klubnichkin et al. 2015; Kochnev et al. 2008; Kotikov 1995; Krasnenkov et al. 1984; Rakheja et al. 1992; Said Al-Milli et al. 2010; Silaev 1972; Wong 2001; Zabavnikov 1975) and others.

Interaction between the tracked running gear and snow have been reflected in the works by (Abol 1951; Bekker 1969; Kochnev et al. 2008; Wong 2009; Zabavnikov 1975) and others.

Review of the contemporary knowledge about interaction between tracked running gear and crushable soil concludes the following.

Today impact of the track width and length on the vehicle drought and qualitative impact of those parameters on the travel resistance are mostly studied.

Out of the great number of formulas developed to determine pressure on soil and travel resistance of the tracked vehicles on different soils none can be used to calculate the above values in case of vehicle's operation on deep virgin snow.

The least examined issue of the process of interaction between the tracked vehicle and the path bed is re-distribution of pressure along the track ground contacting area which occurs under external forces as well as impact of uneven pressure distribution per travel resistance unit.

Considering the above, we deem it necessary to set the following researching tasks:

- to develop a method of analytical determination of optimum displacement of the center of gravity and maximum pressure of the tracked vehicle as well as resistance to its travel over crushable soil;
- to experimentally test the accuracy of the proposed method on the skidding tractors moving over forest virgin snow and obtain required correction factors;
- to analyze the obtained relationships and basing on the analysis provide recommendations for reduction of resistance to travel of the tracked skidding vehicles over virgin snow.

2. Analysis of interaction between tracked running gear of the timber harvesting machine and crushable soil

Movement of the timber harvesting machine over crushable soil is accompanied by rutting that spends work. The amount of this work and caused resistance to travel are determined by maximum pressure transferred by the vehicle to the path bed which is normally much higher that average calculated vehicle pressure on soil.

There are two groups of reasons causing uneven pressure along the length of the track ground contacting area. The first group includes a ratio of side roller pitch and track pitch, ratio of spring arms, track tension. As a result of action of the above factors the vehicle to soil pressure diagram may have either gentle curves or pronounced peaks under rollers and valleys between them. We shall further refer to this type of irregularity as a peak irregularity. As previously noted, the peak irregularity (for this type of soil) is determined exclusively by design concept of the undercarriage.

The second type of irregularity which shall be referred to as a longitudinal irregularity is caused by impact of external forces on the vehicle: load from working elements and travel resistance. It leads to displacement of the diagram center of gravity relative to the vehicle's center of gravity. It is natural that application of external forces cannot change the nature of peak irregularity of the vehicle's pressure on soil but it can significantly influence the value of maximum pressure and thus the travel resistance.

Other things being equal, the vehicle's maximum soil pressure caused by peak irregularity will have the least value in most cases when the center of gravity rests on a vertical being an axis of symmetry of the ground contacting area.

Since the work reviews impact of external forces on re-distribution of pressure along the length of the ground contacting area, then the main attention will be paid to the pressure longitudinal irregularity.

The process of interaction between the track running gear and the crushable soils is quite complex. This is noted by many authors (Bekker 1969; Kotikov 1995; Wong 2009; Zabavnikov 1975). The track under action of vertical and horizontal reactions of the soil gets bent at the front slanting runs and gets bulged at the intervals between support rollers. With that the tracks while in contact with the soil execute complex oscillatory motion and transfer maximum pressure to the path when support rollers pass over tracks. In most cases when research is made for interaction between the track running gear and the crushable soil, it is necessary to review this process with due consideration of all details.

However, when analyzing re-distribution of pressure along the length of the track ground contacting area (longitudinal irregularity), we may use a widely held in the tractor theory assumption that pressure is transferred to soil without peaks under rollers and the suspension of the vehicle's running gear is rigid. This assumption is admissible because it is not the form of the diagram but the position of its center of gravity, i.e. a point of application of the resultant ground reaction which is important for solving this problem.

When reviewing equilibrium of the tracked vehicle at different positions of the center of pressure we obtain relationship between maximum pressure and value of displacement of the center of pressure from the middle of the ground contacting area this or other way:

at $\varepsilon \geq -\frac{1}{6}L$

$$q_{max} = q_0\left(1 + \frac{6|\varepsilon|}{L}\right), \tag{1}$$

at $\varepsilon \leq -\frac{1}{6}L$

$$q_{max} = \frac{4q_0L}{3L-6|\varepsilon|} \tag{2}$$

Value ε is accepted negative at displacement of the center of pressure from the middle of the ground contacting area to the side opposite to the movement direction, and positive in case of reverse displacement. So, relationship between maximum pressure and value of displacement ε of the center of pressure from the center line of the track ground contacting area will be linear until

$$\varepsilon \geq -\frac{1}{6}L$$

Should this inequality is violated, the diagram of pressure under the bottom runs become triangular, and the relationship between q_{max} and ε becomes hyperbolic (formula 2). In other words, displacement of the center of pressure from the middle of the track to the side opposite to the movement direction by the value more than $\frac{1}{6}L$ leads to sharp increase in maximum pressure. This pressure will define the depth of vehicle's landing, work of deformation at rutting and hence resistance to travel.

That is why to provide movement of the tracked timber harvesting machine this or other way it is necessary that the center of pressure gets displaced from the track middle to the side opposite to the movement by the value not exceeding $\frac{1}{6}L$.

Proceeding from these conditions, the maximum displacement of the vehicle's center of gravity forward from the middle of ground contacting area shall be defined by the following equation

$$\varepsilon_{0\,max} = \frac{1}{6}L - h_c tg\beta \tag{3}$$

Determining center of pressure. As it has noted, one of the main criteria of longitudinal irregularity of pressure distribution along the length of the track ground contacting area is a position of the center of pressure, i.e. a point of application of resultant normals to the ground reactions. By knowing the value and point of application of the total normal reactions we can find maximum pressure of the forest vehicle on the soil (with due consideration for peak irregularity), and hence resistance to travel caused by rutting. Besides, in case of necessity we can find soil reaction to support rollers.

It needs to be noted that since modern tracked skidding vehicles has a speed of around 4 km/h during travel with load over crushable surface, then the entire vehicle's interaction with soil is analyzed without consideration of such forces as air environment resistance, inertial force, as well as impact of travel speed on resistance to travel. Those factors have little impact on the reviewed process and they can be neglected.

Let us review a case of irregular straight line movement of the tracked vehicle followed by rutting when it is influenced by hook effort, vertical load from drag and frontal resistance to movement. In a general case a vehicle will move with a certain longitudinal trim β which occurs due to action of external forces.

Sum moment of forces relative to point **A** (Figure 1):

$$\sum M_A = X_{п1}(h_x + Lsin\beta) + X_b[h_x ctg(\psi + \beta) + Lcos\beta] - G_0\left(\frac{L}{2} + \varepsilon_0\right)cos\beta + +G_0 h_c sin\beta + W\left(\frac{L}{2} - \varepsilon\right) -$$
$$- Q\left(\frac{L}{2} - \delta_0\right)cos\beta - P_{кр}\left(\frac{L}{2} - \delta_0\right)\sin(\gamma - \beta) + Qh_{кр}sin\beta + +P_{кр}h_{кр}cos(\gamma - \beta) = 0, \tag{4}$$

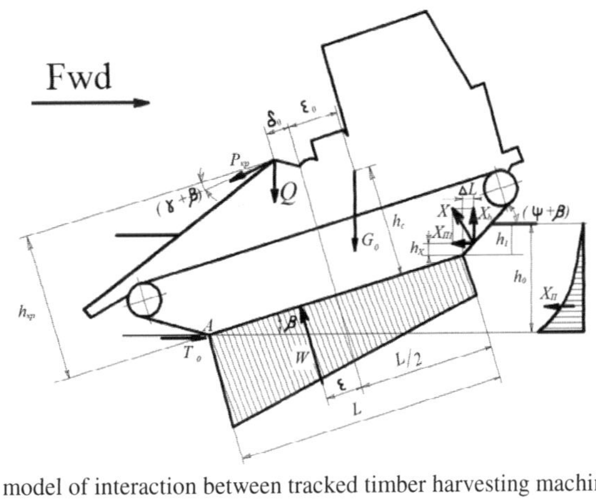

Figure 1. Calculation model of interaction between tracked timber harvesting machine and crushable soil

IOP Conf. Series: Materials Science and Engineering **695** (2019) 012004 doi:10.1088/1757-899X/695/1/012004

Formula (3) suggests that the value of backward displacement of the center of pressure from the track lateral axis stands at

$$\varepsilon = \frac{L}{6}\left(1 - \frac{q_п}{q_0}\right). \tag{5}$$

By substituting value ε into equation (4) and considering that $X_b = X_п ctg(\frac{\psi}{2} + \beta)$, after the transformation we obtain:

$$X_{п1}\left[h_x\left(1 + ctg\left(\frac{\psi}{2} + \beta\right)ctg(\psi + \beta)\right) + L\left(sin\beta + ctg\left(\frac{\psi}{2} + \beta\right)cos\beta - \frac{ctg\left(\frac{\psi}{2}+\beta\right)}{3cos\beta}\right)\right] - -G_0\left[\left(\frac{L}{2} + \varepsilon_0\right)cos\beta - h_c sin\beta - \frac{L}{3cos\beta}\right] + Q\left[\frac{L}{3cos\beta} - \left(\frac{L}{2} - \delta_0\right)cos\beta + h_{кр}sin\beta\right] + P_{кр}\left[\frac{Lsin\gamma}{3cos\beta} + +h_{кр}\cos(\gamma - \beta) - \left(\frac{L}{2} - \delta_0\right)\sin(\gamma - \beta) + \frac{bL^2}{3}q_п = 0. \tag{6}$$

Modern tracked timber harvesting machines while moving over soft soil and snow in normal operation conditions have a longitudinal trim angle of β not exceeding 8°. Horizon tilting angle of of the cable assembly γ also lies within 8° to 10°. This allows for some assumptions without great damage to accuracy:

$$cos\beta = 1;\ \cos(\gamma - \beta) = 1;\ \ sin\beta = tg\beta;\ \ sin^2\beta = 0.\ \ \sin(\gamma - \beta) = 0.$$

Considering those assumptions, after transformation of equation (6) we obtain:

$$(G_0h_c + Qh_{кр})sin\beta + X_{п1}\left[\frac{h_x(1+cos\psi)}{sin^2\psi+(sin^2\psi+sin\psi)\cdot sin\beta} + \frac{L(2cos\frac{\psi}{2}+sin\frac{\psi}{2}\cdot sin\beta)}{3sin\frac{\psi}{2}+3\ cos\frac{\psi}{2}\cdot sin\beta}\right] + \frac{bL^2}{3}q_п - -G_0\left(\frac{L}{6} + \varepsilon_0\right) - Q\left(\frac{L}{6} - \delta_0\right) + P_{кр}\left[\frac{Lsin\gamma}{3} + h_{кр}\right] = 0. \tag{7}$$

In this equation values $sin\beta$, $X_{п1}$ are $h_x$variable and depend on the properties of the crushable soil and type of undercarriage. These values can be expressed in terms of conditional value of pressure under the front rollers $q_п$.

Figure 1 suggests that

$$sin\beta = \frac{h_0 - h_1}{L}, \tag{8}$$

where h_0 and h_1 - depth of track under rear and front rollers respectively.

Values $X_{п1}$ and h_x are revealed in function q_n after review of a diagram of interaction of the forward slanting section of a track with crushable soil

$$X_{п1} = 2b\int_0^{q_п} f_2(q)dq, \tag{9}$$

$$h_x = \frac{0{,}5\int_0^{q_п}[f_2(q)]^2dq}{\int_0^{q_п} f_2(q)dq}. \tag{10}$$

By substituting values of formulas (9) and (10) into equation (7) we obtain after transformations;

$$(G_0h_0 + Qh_{кр})sin\beta + \frac{2}{3}bL\left[\frac{2cos\frac{\psi}{2}+sin\frac{\psi}{2}\cdot sin\beta}{sin\frac{\psi}{2}+cos\frac{\psi}{2}\cdot sin\beta}\right]\int_0^{q_п} f_2(q)dq + \frac{b(1+cos\psi)\int_0^{q_п}[f_2(q)]^2dq}{sin2\psi+(sin2\psi+sin\psi)sin\beta} + +\frac{bL^2}{3}q_п - G_0\left(\frac{L}{6} + \varepsilon_0\right) - Q\left(\frac{L}{6} - \delta_0\right) + P_{кр}\left[\frac{Lsin\gamma}{3} + h_{кр}\right] = 0. \tag{11}$$

Equation (11) is solved relative to value $q_п$- one of the trapezoid sides - pressure diagram. If we consider that the second side of the trapezoid is $q_k = 2q_0 - q_п$, then we can take that the position of the center of pressure is known to us. This equation is expressed in a general view. To obtain concrete results it is necessary to substitute required parameters of a tracked tractor and soil properties into it. The properties of the crushable soil are expressed by relationship $h = f(q)$between deformation and pressure which can be found experimentally.

Resistance to travel of the timber harvesting machine caused by rutting

The timber harvesting machine while moving over crushable soil and making track faces certain resistance to travel. Resistance due to soil crushing by tractor's tracks i.e. rutting constitutes the major part of this resistance. The value of this resistance is directly proportional to the amount of the pressed soil.

For the sake of simplicity the resistance due to rutting can be viewed as two forces. One acts on the front slanting runs of the track, the other acts on the ground contacting area.

In lack of so called "bulldozer" effect, i.e. soil pushing by tracks, the resistance faced by the front slanting runs of the tracks is expressed by the equation (9):

$$X_{\Pi 1} = 2b \int_0^{q_\Pi} f_2(q)dq. \tag{12}$$

Similarly the resistance acting on the ground contacting area can be expressed as follows:

$$X_{\Pi 2} = 2b \int_{q_\Pi}^{q_{max}} f_2(q)dq. \tag{13}$$

The total resistance to travel due to rutting will be composed of two components

$$X_\Pi = X_{\Pi 1} + X_{\Pi 2} = 2b \int_0^{q_{max}} f_2(q)dq. \tag{14}$$

3. Experimental research method

The experimental research is aimed at establishing the relationships which characterize the process of rutting and equations required for and deduced in the previous section. In particular, it was necessary to establish relationship between the track landing depth and pressure, the coefficient of pressure concentration under roller, relationship between resistance to travel and position of center of gravity and center of pressure, and to obtain experimental material to check the method of calculation of pressure re-distribution along the length of the track ground contacting area.

All experiments were carried out at the tracked tractor TT-4M during its interaction with deep forest virgin snow as a form of crushable soil.

To maintain experimental integrity the tests were done on forest glades with even surface. The snow depth varied from 0.6 to 0.9 m. Measures were made in Murashinsky timber industry enterprise of Kirov oblast, Russian Federation

Numerous measurements of scow density made at different sites and in different years have demonstrated that average values of forest snow density remained within 0.23 to 0.27 g/cm^3. This allowed for a conclusion that the properties of forest snow fell at stable negative temperatures and not subjected to thaw change insignificantly and those changes have no impact on the experiment results.

During the experiments the recording equipment was placed inside a cab of the timber harvesting tractor. To check the calculation method it was necessary to measure normal pressure and resistance to travel at different positions of the center of pressure of the tested tractor.

To simulate forward displacement of the center of pressure the tractor was equipped with two-meter arms where a 1-ton load was secured. The arms provided for three positions of the load. Backward displacement of the center of pressure was provided by securing a load on board in three positions as well. Besides the tractor was subjected to significant backward displacement of the center of pressure as of roundwood bundle and dragging another tractor by a hoist cable with strain gauge.

Measurements were made with a pressure transmitter, a snow track landing transmitter, strain gauge system to measure torque on driving sprockets.

A special test bench was used to determine tractor's center of gravity.

4. Results of experimental research

As a result of the pressure and landing oscillograms' processing we obtained relationships between specific pressure and depth of landing of tractor TT-4M track landing into snow 0.8 m deep.

$$q = \frac{h}{200 - 2{,}5h}. \tag{15}$$

For snow 0.6 m deep this relationship is as follows

$$q = e^{0{,}1h - 3{,}62} \tag{16}$$

Relationships (15) and (16) have been obtained while reviewing interaction between snow and front slanting section of the track to the first support roller. Hence, these relationships are the function of snow properties and track width. Other factors, including type of suspension, weight parameters, tilting angle of front sections (in lack of or insignificant value of snow pushing) have no impact on them.

This brings to a conclusion that relationships between deformation and pressure obtained for tractor TT-4M, which track width is equal to 50 cm may be applied to calculation for any tracked vehicles which feature track width range within 50 cm.

By substituting values of function by equation (15) into formula (3) and considering that tgβ can be replaced with sinβ without great mistake, we obtain maximum forward displacement of the tractor's center of gravity (for snow H=0.8 m)

$$\varepsilon_{omax} = \frac{L}{6} - \frac{400h_c q_0}{(5q_0+1)L} \tag{17}$$

Value of average specific pressure is expressed as follows

$$q_0 = \frac{G_0}{2bL}$$

After revealing values $X_{П1}$ and h_x with due consideration of dependences (15) and (16) we obtain expressions for finding values $q_П$ for snow of:

H=0.8 m

$$(G_0 h_c + Qh_{KP})sin\beta + 21{,}3bL\left(\frac{2cos\frac{\psi}{2}+sin\frac{\psi}{2}sin\beta}{sin\frac{\psi}{2}+cos\frac{\psi}{2}sin\beta}\right)\cdot\left[\ln(2{,}5q_п+1) - \frac{2{,}5q_п}{2{,}5q_п+1}\right] + \frac{136b(1+cos\psi)\left(\frac{2{,}5q_п}{2{,}5q_п+1}\right)^3}{sin^2\psi+(sin2\psi+sin\psi)sin\beta} + \frac{bL^2}{3}q_п - Q\left(\frac{L}{6}-\delta_0\right) + P_{кр}\left(\frac{Lsin\gamma}{3}+h_{кр}\right) - G_0\left(\frac{L}{6}+\varepsilon_0\right) = 0; \tag{18}$$

H=0.6 m

$$(G_0 h_c + Qh_{KP})sin\beta + 21{,}3bL\left(\frac{2cos\frac{\psi}{2}+sin\frac{\psi}{2}sin\beta}{sin\frac{\psi}{2}+cos\frac{\psi}{2}sin\beta}\right)\cdot q_п + \frac{100b(1+cos\psi)q_п}{sin^2\psi+(sin2\psi+sin\psi)sin\beta} + \frac{bL^2}{3}q_п - Q\left(\frac{L}{6}-\delta_0\right) + P_{кр}\left(\frac{Lsin\gamma}{3}+h_{кр}\right) - G_0\left(\frac{L}{6}+\varepsilon_0\right) = 0; \tag{19}$$

In those equations

At H=0.8 m

$$sin\beta = \frac{32(G_0+Q-2bLq_п)}{(0{,}4+q_п)[G_0+Q+bL(0{,}4-q_п)]L} \tag{20}$$

At H=0.6 m

$$sin\beta = \frac{10}{L}\ln\left(\frac{G_0+Q}{bLq_п} - 1\right) \tag{21}$$

The tracked vehicles moving over snow and experiencing no hook load normally feature insignificant longitudinal trim angle (1 – 2°max). This angle has little impact on calculation results and they can be neglected.

Then for vehicles with similar load diagram equations (18) and (19) will take the following form:

H=0.8 m

$$42{,}7bL\ ctg\frac{\psi}{2}\left[\ln(2{,}5q_п+1) - \frac{2{,}5q_п}{2{,}5q_п+1}\right] + \frac{136b}{1-cos\psi}\left(\frac{2{,}5q_п}{2{,}5q_п+1}\right)^3 + \frac{bL^2}{3}q_п = G_0\left(\frac{L}{6}+\varepsilon_0\right) \tag{22}$$

H=0.6 m

$$q_п = \frac{G_0\left(\frac{L}{6}+\varepsilon_0\right)}{13{,}3bL\ ctg\frac{\psi}{2} + \frac{100b}{1-cos\psi} + \frac{bL^2}{3}} \tag{23}$$

Basing on formula (14) with due consideration of relationships (15) and (16) we obtain expressions for finding travel resistance due to rutting:

H=0.8 m

$$X_п = 64b\left[\ln(2{,}5q_{max}+1) - \frac{2{,}5q_{max}}{2{,}5q_{max}+1}\right] \tag{24}$$

H=0.6 m

$$X_п = 20bq_{max} \tag{25}$$

Determining pressure concentration coefficient. *The* oscillograms of normal pressure recorded during movement of the skidding tractor over snow clearly show the peaks under the rollers and valleys between rollers where pressure drops almost to zero. This concentration (peak irregularity) of pressure under rollers depends on physical and mechanical properties and depth of snow, pitch of support rollers, pitch of track links, type of suspension, towing and other factors. For each specific case it can be evaluated by pressure concentration coefficient k, equal to ration of average pressure during peaks to average calculated pressure q_0

$$k = \frac{\Sigma q_i}{n q_0} \tag{26}$$

where Σq_i – a sum of pressures in peaks under rollers;

n – number of rollers on one side of the tractor.

Experiments have established that during movement of the tracked vehicle with ration of roller pitch to track link pitch of $\lambda = 4.5$ (TT-4M and LP-18K) over snow 0.8 m deep, coefficient k is equal to 1.95, during those vehicles' movement over snow 0.6 m deep - 2.45. This suggests that due to peak irregularity the true pressure exceeds calculated one 2 - 2.5 times.

Calculating maximum pressure. Finding a coefficient of pressure concentration allowed for consideration of peak irregularity during calculation of maximum pressure. For calculating maximum pressure it is first necessary to determine value q_n by equations (18) or (19) depending on snow depth. If value q_n exceeds average calculated pressure q_0, then maximum pressure shall be described in terms of the following relationship

$$q_{max} = k \cdot q_n \tag{27}$$

If value q_n is less than q_0, then maximum pressure expression shall take the following form

$$q_{max} = k \cdot (2q_0 - q_n) \tag{28}$$

Comparing experimental values of maximum pressure with calculated one by formulas (27) and (28) demonstrates that maximum deviation of the calculated values from the experimental data does not exceed 5%.

Relationship between resistance to travel and maximum pressure. As a result of processing of oscillograms of recording torque of driving sprockets we obtain resultant circumferential force P_0. The total resistance to vehicle's travel $P_{xч}$ is determined as a difference between resultant circumferential force P_0 on driving sprockets and drawbar force $P_{кр}$

$$P_{xч} = P_0 - P_{кр} \tag{29}$$

In its turn the total resistance to travel $P_{xч}$ is composed of resistance to travel due to rutting X_n, internal losses in the undercarriage P_t and losses due to rubbing of the vehicle's bottom and protruding parts against snow, entry of snow inside a track, increased tension of tracks and other factors.

$$P_{xч} = X_П - P_T + \Delta P \tag{30}$$

Going to the coefficients, we can write

$$f_{xч} = f_П - f_T + \Delta f \tag{31}$$

Isolating all losses which comprise value ΔP, from the total resistance to travel is a complex task. This value depends on the type of undercarriage, depth and properties of snow as well as load on the vehicle. However, for certain types of undercarriage, snow depth and specific load on a tractor value ΔP will be more or less constant. For this, to obtain an experimental value of force X_n, losses in undercarriage P_t were subtracted from total resistance to travel $P_{xч}$.

Basing on the experimental data a diagram (Fig. 2) of relationship between resistance to travel of tractor TT-4M and maximum pressure on snow was built. Analytical curve 1 is plotted by equation (24). Experimental values points of value X_n do not match this curve and lie above it. This can be explained by that when calculating value X_n no consideration was made for value ΔP, which was found significant for snow 0.8 m deep. The major part of this resistance belongs to bottom rubbing against snow as with snow depth of 0.8 m the track depth of the loaded tractor TT-4M exceeds clearance by 0.10 —0.15 m.

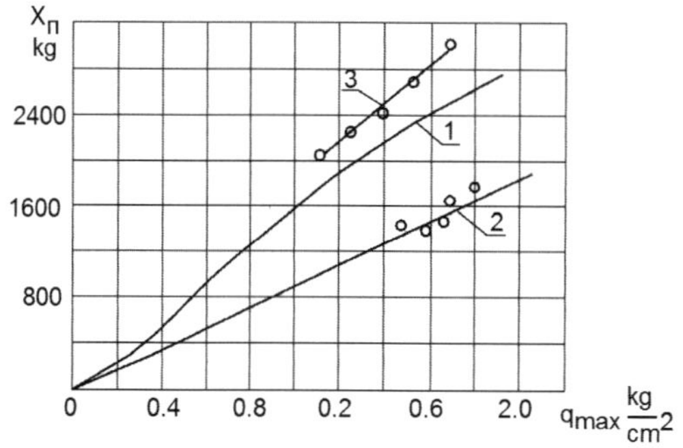

Figure 2. Relationship between resistance to travel of tractor TT-4M and maximum pressure:
1 – curve by equation (18) – for snow H = 0.8 m;
2 – curve by equation (19) – for snow H = 0.6 m;
3– experimental curve for snow H = 0.8 m.

Curve 2 (see Fig. 2) is plotted by formula (25) for snow depth of 0.6 *m*. As seen, all experimental points slightly deviate from the curve. In fact when tractor TT-4M interacts with snow 0.6 *m* deep, its bottom does not contact the snow surface, the rest factors have no major impact on resistance to travel. This suggests that if the bottom does not press snow the value of additional losses ΔP is insignificant and it can be easily neglected.

Ordinates of curve 1 (see Fig. 2) are inevitable part of resistance to travel caused by rutting by tracks which a vehicle is subjected to when on the snow at this or that maximum pressure. Curve 3 connecting experimental points of resistance X_n, shows that actual values of this resistance are above calculated ones by 10 – 14%. This increase of resistance mainly caused by snow pressing by a vehicle bottom is a reserve of reduction of frontal resistance which may be implemented for an identical vehicle on the snow if its bottom does not contact the snow surface.

On the other hand, curves 1 and 2 offer an opportunity to visualize the effect which can be gained by eliminating the peak and longitudinal irregularities of pressure. Thus, under ideally uniform distribution of pressure tractor TT-4M moving with a load shall transfer to snow a pressure $q \approx 0{,}5$ kg/cm^2.

Relationship between resistance to travel and center of pressure displacement. When developing a curve of relationship between resistance to travel and displacement of center of pressure, it is more convenient to describe resistance to vehicle's travel in terms of coefficient of travel resistance, and displacement of center of gravity and a center of pressure , in terms of fractions of length of ground contacting area. Such diagram for snow depth of 0.6 *m* is given in Fig. 3*a*. Experimental points of resistance coefficient f_n coincide with a theoretical curve satisfactory (deviation does not exceed 6%).

Fig. 3*b* shows the same diagram for snow depth of 0.8 *m*. All experimental points lie above the theoretical curve. As it has been already noted, this is a result of interaction of the tractor with snow as the track landing depth is more than its clearance by 140 mm.

However, as it is seen on the diagram, with center of pressure displacement back from the middle of the ground contacting area by value $\varepsilon \geq 0.02L$ the difference in ordinates between theoretical and experimental curves is almost constant and stands at 14%. in average With center of pressure displacement back by value $\varepsilon < 0.02L$ and forward from the middle of the track, this difference rises sharply and reaches 28%.

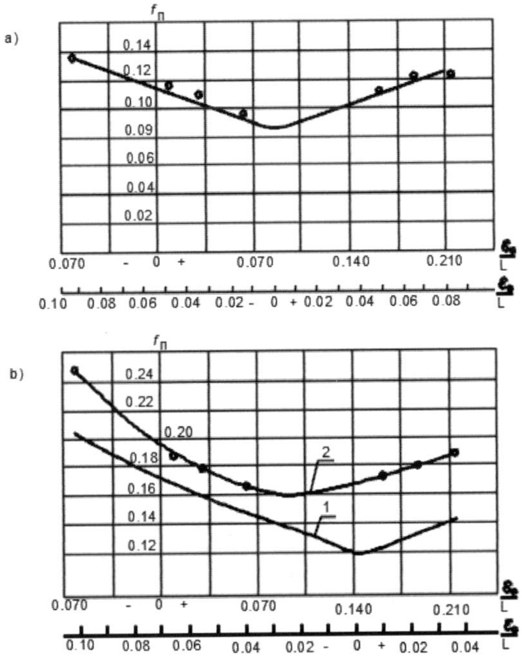

a) H = 0.6 m;
b) H = 0.8 m.
1 — theoretical curve; 2 — a curve connecting experimental points.
Figure 3. Relationship between travel resistance coefficient and displacement of center of gravity and center of pressure of tractor TT-4M for snow depth of:

This can be explained by that with displacement of center of pressure back from any point (in this case it is 0.02 *L*)a vehicle experiences longitudinal trim. With that the bottom compacts the snow, rubs against its surface thus creating almost constant additional resistance. Displacement of center of pressure forward from the given point leads to practical disappearance of a trim as a result of which the front head part of the bottom pushes the upper layer of snow thus forming a snowbank which sharply increases resistance to travel.

In normal operation a center of pressure of the skidding tractor is displaced backward by a value exceeding 0.02 *L*. That is why to estimate passability of the modern skidding vehicles which clearance is about 500 mm, over virgin snow 0.8 *m* deep, it is necessary to increase estimated value *Xn,* obtained against expression (18), by 14%.

The performed analysis and experience of the skidding vehicle's operation in deep virgin snow demonstrate that snow pressing by a bottom causes significant reduction of their passability. In addition to increase of resistance to travel due to bottom rubbing against snow, the vehicle's adhesion weight reduces and this reduction reaches 19% in case of tractor TT-4M. In case of track's skidding the vehicle's landing depth increases and a vehicle "sits" on its belly.

That is why the vehicles designed for operation in deep snow shall have a clearance which value is more or equal to track depth made in snow during movement under full load.

5. Conclusion

The proposed method of determination of the value of maximum forward displacement of the center gravity, maximum pressure and resistance to travel of the tracked vehicle is proved experimentally and can be used in designing new and evaluating existing vehicles from the point of their passability in deep virgin snow.

Assessment of the peak irregularity of pressure transfer along the length of the track ground contacting area can be checked with use of pressure concentration coefficient *k* equal to ration of the average peaks pressure i to average estimated pressure of the vehicle. This coefficient depends on ration λ of the support roller pitch to track link pitch and snow depth and properties.

For the skidding vehicles made by Altay tractor plant with value $\lambda = 4.5$ coefficient *k* is equal to 1.95 for snow depth of 0.8 m, and 2.45 for snow depth of 0.6 m.

When a modern tracked skidding vehicle moves with a load over forest virgin snow 0.8 – 0.9 m deep, the value of tracks' landing into snow exceeds clearance by 0.10 – 0.15 m. The bottom rubbing against snow leads to sharp increase of resistance to travel, especially when a center of pressure is displaced forward from the track middle, due to snow pushing by the tractor's bottom front part.

Analysis of calculated relationships and experimental data demonstrates that:

a) leveling of the pressure diagram by eliminating peak irregularity reduces resistance to travel of the modern skidding tractors with snow depth of 0.8 – 0.9 m twofold, and with snow depth of 0.6 m, in 2.5 times;

b) eliminating longitudinal irregularity (displacement of a center of pressure with lateral axis of symmetry of the track ground contacting area) may reduce resistance to travel in 1.5 – 1.8 times.

Increasing the length of the ground contacting area significantly increases capacity of the tracked skidding vehicles during operation on virgin snow.

The calculations found that increasing the length of the ground contacting area of tractor TT-4M by 10% leads to increase of maximum truck load by 28%, and increasing the length of the ground contacting area and clearance by 10% provides increase of truck load by 44%.

The proposed general analytical equations may be used for calculation of interaction of variable types of tracked undercarriage with different crushable soils. To do so it is necessary to expand the given equations using relationships between deformation and pressure obtained under the proposed method.

Nomenclature

$q_0 = \frac{G_0}{2bL}$	average specific pressure of the track ground contacting area on the soil	[kg/cm^2]
b	width of a track	[m]
G_0	vehicle weight	[N]
L	length of track ground contacting area	[m]
h_c	vertical coordinate of the center of gravity	-
β	angle of vehicles' longitudinal trim	[°]
Q	vertical load by drag	[kN]
$P_{кр}$	hook load	[kN]
$X_{п1}$ and X_b	horizontal and vertical components of resultant soil reaction to frontal area of a track respectively	-
W	resultant normals to ground reactions	-
$h_{кр}$	height of application of hook force	-
β	angle of longitudinal trim	[°]
γ	angle of inclination to cable assembly horizon	[°]

Acknowledgements
Research financed by the Ministry of Science and Higher Education of the Russian Federation project number № 9.8996.2017/8.9.

References

[1] Abol I.P., 1951. Investigation of the influence of the nature of snow fibers on the tractive qualities of the KT-12 skidder. Dis. Cand. technical sciences. Moscow. Pages 16.
[2] Anisimov, G.M., 1975. Conditions of operation and loading of transmission of tractor skidding. Moscow. Forest industry. Pages 166.
[3] Bekker, M.G., 1969. Introduction to Terrain-Vehicle Systems. The University of Michigan Press, Ann Arbor.
[4] Bodin, A., 1999. Development of a tracked vehicle to study the influence of vehicle parameters on tractive performance in soft terrain. Journal of Terramechanics, Vol. 36, Issue 3, 167–181.
[5] Dhir, A. and Sankar, S., 1994. Analytical track models for ride dynamic simulation of tracked vehicles. Journal of Terramechanics, Vol. 31, Issue 2, 107–138.
[6] Dmitriev, A.A. Savochkin, V.A., 1993. Statistical dynamics of transport and traction tracked vehicles. Moscow. Mashinostroenie, Pages 320.
[7] Cambi, M., Certini G., Neri F., Marchi, E., 2015. The impact of heavy traffic on forest soils. Forest Ecology and Management. Vol. 338, 124-138.
[8] Goberman, L.A., Goberman V.A., Ksenevich, I.P., 2003. Ground trailer transport systems. Technical and economic bases of designing machines and processes. Methodological aspects of project activities and decision making. Volume 3. Encyclopedia. Engineering, Moscow. Pages 788.
[9] Horn, R., Vossbrink, J., Peth, S., Becker, S., 2007. Impact of modern forest vehicles on soil physical properties. Forest Ecology and Management. Vol. 248, Issues 1–2, 56-63.
[10] Klubnichkin, V., Klubnichkin, E., 2015. Research of kinematics and dynamics of tracked timber harvesting vehicle running gear. Proceedings of the 48th International Symposium on Forestry Mechanization, 315–320, 04–08 October 2015, Linz (Austria), ISBN 978-3-900932-29-9.
[11] Kochnev A.M., Anisimov, G.M., 2008. Tests of harvesting machines. SPb.: SPbGLTA, Pages 488.

[12] Kotikov, V.M.,1995. Impact harvesting machines on forest soils. Dissertation of the doctor of technical sciences. Moscow state forest university, Pages 214.
[13] Krasnenkov, V.I., Egorkin V., Kharitonov S.A., 1984. Mathematical model of curvilinear motion of tracked vehicle transport on non-deformable base. Izv. Engineering, Volume 5, 81–85.
[14] Rakheja, S., Afonso, M.F.R. and Sankar S., 1992. Dynamic analysis of tracked vehicles with trailing arm suspension and assessment of ride vibrations. International Journal of Vehicle Design, Vol.13, No.1.
[15] Said Al-Milli, Lakmal D. Seneviratne, Kaspar Althoefer., 2010. Track–terrain modelling and traversability prediction for tracked vehicles on soft terrain. Journal of Terramechanics, Volume 47, Issue 3, 151–160.
[16] Silaev, A.A., 1972. Spectral theory of suspension of transport vehicles. Mechanical Engineering, Moscow. Pages 192.
[17] Wong, J.Y., 2001. Theory of Ground Vehicles, third ed. John Wiley & Sons, NY, ISBN 0-471-35461-9.
[18] Wong, J.Y., 2009. Terramechanics and Off-Road Vehicle Engineering, 2nd Edition. Pages 488.
[19] Zabavnikov N.A., 1975. Fundamentals of the theory of transport tracked vehicles. Moscow. Mashinostroenie, Pages 448.

Optimization of the internal damping rubber elements of the roller wheel of tracked system

S A Korostelev[1] **and A F Verbilov**[2]

[1] Department of land transport-technological systems, Polzunov Altai State Technical University, Barnaul, Altai Region, Russia
[2] Firearms and Technical Training Department, Barnaul Law Institute of the Ministry of Internal Affairs of Russia, Barnaul, Altai Region, Russia

E-mail: korsan73@mail.ru

Abstract. The article presents the formulation and solution of the problem of optimization of the internal damping rubber elements of the roller wheel of tracked system. The optimal design search is based on the results of calculating the stress-strain state of the rubber element caused by the assembly of the roller and operational loads. As an objective function, the specific energy of deformation of the rubber was chosen. The geometric parameters of the rubber element of the roller wheel in the initial state and after installation in the structure are taken as variable parameters. Explicit and functional constraints in the form of inequations are imposed on the variable parameters. As a result of solving the optimal design problem, the shape and design parameters of the internal damping rubber elements of the track roller of the tracked system were obtained, which made it possible to reduce the specific strain energy during radial loading by 1.85 times

1. Introduction

Expanding the scope of tracked vehicles, increasing the requirements for the durability of elements of a tracked mover while increasing specific loads and the desire to reduce metal consumption [1, 2] forces engineers to use new solutions in the design of track rollers.

The increase in the speeds of the high-speed tracked vehicles and the energy saturation of agricultural tractors led to an increase in the dynamic loads acting on the elements of the tracked mover. The high level of dynamic stresses affects the strength, reliability and durability of the elements of the tracked system. Reducing dynamic loads affecting the elements of the traked mover is an important engineering problem [1, 3 - 5].

2. Features of the use of rubber elements in the constructions of roller wheels

One of the possible designs of the roller wheel, which allows to reduce dynamic loads, is a construction, in which the elastic connection between the hub and the metal bandage is carried out using internal damping rubber elements [1, 2, 6].

In roller wheels with internal rubber elements, rubber works in favorable conditions [2]. Namely, most of it is protected from aggressive environments, rubber elements perceive compression and shear deformations, rubber is protected from direct exposure to foreign objects penetrating between the track and the roller wheel. All this allows to significantly increase the durability of the rubber element. Another important advantage of designs with internal rubber elements is their maintainability. In such case, both the rubber elements and the bandage are the least durable, the replacement of which during the repair of the caterpillar mover, for some designs, is possible without dismantling the entire roller.

The geometric shape of the power rubber elements determines their characteristics and, first of all, elastic and compensation properties. Suffice it to say that the stiffness in compression, torsion or shear, the values of permissible deformations of individual types of elements of the same overall size, can differ by an order of magnitude. The geometric shape has a significant effect on the damping ability of rubber elements.

When designing power rubber elements of a tracked system, one of the main issues is the question of their durability. The existing variety of design of power rubber elements does not allow focusing only on experimental research data, since their implementation are expensive and time-consuming.

To assess the durability of the rubber elements of the roller wheel of the tracked system, it is advisable to use calculation methods based on the application of the criteria for fatigue strength of rubber. Fatigue strength criteria are formed taking into account the influence of the stress-strain state of rubber elements under static and dynamic loads, temperature, physical and mechanical properties of specific rubber grades.

The process of destruction of the internal damping rubber elements of the roller wheel is subject to the general laws characteristic of all materials: the process of volumetric destruction proceeds in two stages and is accompanied by a period of accumulation of damage and subsequent global destruction of the element; the destruction process is unidirectional and statistical in nature. The destruction of rubber elements during cyclic deformation is accompanied by self-heating, which is due to the viscoelastic properties of the material. Hysteresis phenomena during cyclic deformation of rubber elements in combination with low thermal conductivity require taking into account the effect of temperature on their fracture process.

Power rubber elements subjected to intense cyclic deformation are destroyed by three main damage mechanisms [7]: thermomechanical failure, fatigue failure and mixed failure.

In the case of fatigue failure during the entire period of operation of the rubber element, its temperature remains at a level significantly less than value for a given rubber grade. Fatigue failure begins with the appearance of fatigue microcracks due to the influence of cyclic deformation and subsequently turn into main cracks until the complete destruction of the element.

To assess the fatigue endurance of rubbers from mechanical stress under complex loading conditions, when the sample is subjected to both dynamic and static deformations, dependencies based on the energy criterion are used. The study of the dependence of the fatigue life of rubber using the energy criterion was considered in [7 - 9]. In these works, on the basis of experimental data, it is proved that the specific strain energy is the most accurate and universal criterion for fatigue strength.

Thus, in order to assess the workability of the internal damping rubber elements of the of the roller wheel of a particular design, it is necessary to know their stress-strain state caused by the assembly of the studied unit as well as the stresses, deformations, and temperatures occurring in the body of the rubber element during subsequent operation. In addition, it is necessary to know the stiffness characteristics of the rubber elements and be able to evaluate their influence on the dynamic loads occurring in the elements of the tracked system in order to determine the feasibility of using rubber elements.

3. Theoretical approach for assessing the stress-strain state of the inner damping rubber element of the roller wheel

The phenomenological approach is used to describe the elastic properties of elastomers. It allows to describe the dependences between strains and stresses observed in experiments[10–11].

Nonlinear relationships between stresses and strains in rubber are described using the elastic potential, which is a function of the invariants of the Cauchy strain tensor. The simplest form of elastic potential is the Treloar potential [10, 11]:

$$W = \frac{1}{2}\mu(I_1 - 3), \tag{1}$$

where μ is the shear modulus of rubber.

The Treloar potential allows to describe the properties of a perfectly elastic material in the medium strain region. In this work, the elastic properties of rubber are described by the Treloar elastic potential. Given the dependence (1), the potential energy functional of the system for an incompressible material is written as [12]

$$\Pi = \iiint_v \left[\frac{\mu}{2}(I_1 - 3) + p(\tau - 1)\right] dv - \iint_s \vec{F}\vec{u}ds, \tag{2}$$

where s is the surface of the undeformed volume v; p is the hydrostatic pressure function; $\vec{F}$, $\vec{u}$ respectively are the vector of forces and the vector of displacements at the boundary of the area where external forces are specified.

Internal damping rubber and rubber-metal elements of track roller wheel are bodies of revolution. In this regard, the problem of the stress-strain state is solved in a cylindrical coordinate system.

During assembly, rubber experiences large deformations, therefore, the algorithm for calculating rubber elements is based on the basic relations of the nonlinear theory of elasticity of an incompressible material, namely, the theory of superposition of small elastic deformations on equilibrium finite ones.

In accordance with the algorithm of the method proposed in the works of E. Lavendell the continuous deformation of the rubber element is replaced by a step-by-step procedure. At each step of solving the nonlinear problem, linearized relations of the nonlinear theory of elasticity are applied. Thus, at each step, the linear problem is solved, but the stress-strain state that occurs in the rubber element as a result of deformation at the previous deformation steps is taken into account.

The variation in the increment of potential energy at each step of deformation is determined by the equation [12]:

$$\delta(\Delta\Pi) = \delta\Big(\int_v \overline{W} dv - \frac{1}{2}\int_s \Delta\vec{F} \cdot \Delta\vec{u} ds \Big), \tag{3}$$

where $\overline{W} = \frac{1}{2}\frac{\partial^2}{\partial\eta^2}[W + (P-1)(\tau-1)]$; W is the Treloar elastic potential; η is the small parameter; $\Delta\vec{F}$, $\Delta\vec{u}$ are accordingly, the increment of the force vector and the displacement vector at the boundary of the area where external forces are specified.

The above equations make it possible to determine the stress-strain state of rubber elements both caused by assembly and during secondary loading caused by subsequent operation.

4. Example of the stress-strain state assessment of the internal damping rubber element of the roller wheel

In figure 1 a design variant of the roller wheel [12] with internal cushioning in which the rubber elements operate in shear is presented. Roller wheels with ring rubber elements have a large radial and low axial flexibility, which reduces dynamic loads in the area of contact with the treadmill of the link and eliminates the shift of the bandage during movement of tracked vehicle.

When assembling the roller, the rubber elements are preliminarily compressed in the axial direction, which increases their fatigue strength, and the uniform load of the entire cross section of the ring during shearing, combined with its simple configuration, significantly reduces the probability of the stress concentrators. The design of the roller wheels of this group allows additional tightening of the rubber elements during operation. In addition, structurally, it is possible to ensure the limitation of radial deformation, which allows protecting the rubber elements from overload.

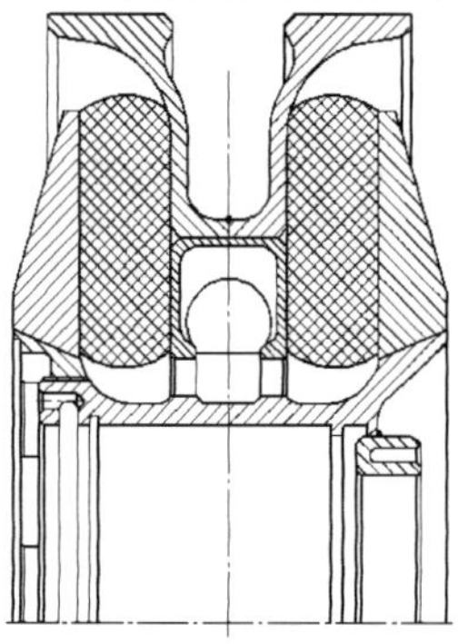

Figure 1. The design of the track roller with internal shock absorption

The structural parameters of the internal damping rubber elements of the roller wheel before and after assembly, together with the loads acting on the roller during operation, determine their stress-strain state. In turn, the stress-strain state of the rubber elements in combination with the loading modes of the roller wheels determines the durability of the rubber elements.

The solution to the problem of determining the stress-strain state after assembly caused by compression in the axial direction is carried out in a cylindrical coordinate system, an axisymmetric problem is considered, the rubber element section is discretized by isoparametric finite elements with eight nodes and non-linear approximation of displacements. Subsequent loading by radial force, the problem is considered in a cylindrical coordinate system and volumetric isoparametric finite elements with twenty nodes are used to solve it.

As a result of the calculation, the distribution fields of the components of strain tensors, stresses, and specific strain energy over the volume of the rubber element of internal depreciation were obtained (Figures 2-4).

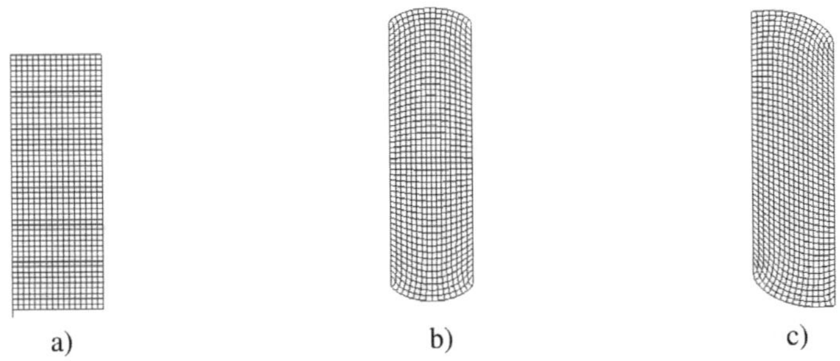

Figure 2. Cross-section of the rubber element of the track roller: a - mesh of finite elements in a free state; b - finite element mesh after compression in the axial direction; c - finite element mesh after applying a radial load to the support bandage

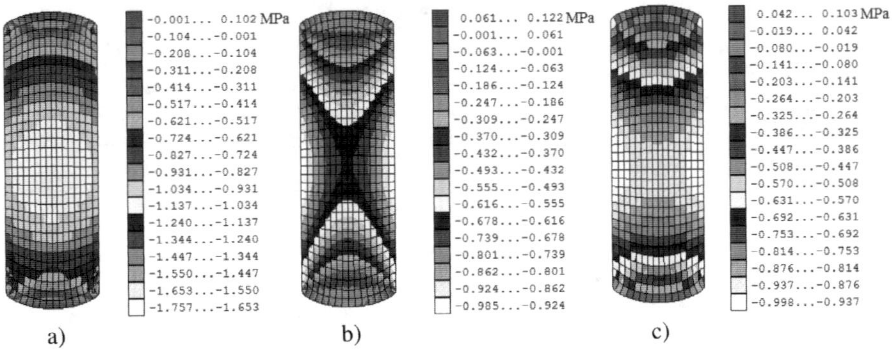

Figure 3. Distribution of normal stresses over the cross section of a rubber element:
a - σ_Z; b – σ_r; c - σ_θ

Figure 4 presents patterns of the distribution of shear stresses caused by assembly deformations, and the specific strain energy caused by subsequent loading by radial force.

The maximum values of shear stresses caused by the assembly for the studied design are valid in areas 1 and 4, 2 and 3 (Figure 4a) and are 0.696 MPa. During secondary loading with the radial force of maximum values, the specific strain energy reaches in areas 1 and 4, 2 and 3 (Figures 4b, 4c). In area 1, its value reaches 100 kJ / m3, and in areas 3 and 4 it exceeds 76 kJ / m3.

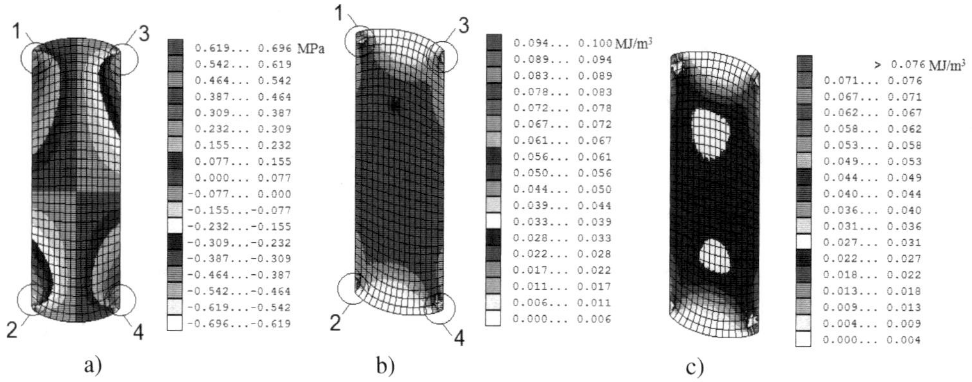

Figure 4. Distribution of shear stresses (a) and specific energy deformations (b, c) over the cross section of the rubber element caused by:
a – assembling; b,c – radial force.

Thus, the tangential stresses reach their maximum values both during assembly and during secondary loading in the same regions, i.e. at the extreme contact points of the rubber element with the surfaces of the rim and the hub of the support roller, therefore, the beginning of fatigue failure for elements of this design will be observed in these areas.

5. Proposal approach of optimal design of rubber elements of roller wheels to increase their durability

The formulation of the problem of optimal design of the internal damping rubber element on the example of the roller wheel shown in figure 1 will be considered. As the objective function, the maximum value of the specific strain energy during secondary loading with radial force will be chosen.

Variational parameters x_j are: the coordinates of the nodal points of the cross section, which determine the shape of the rubber element and geometric dimensions prior to assembly in the initial undeformed state (Figure 3, b); the value of compression acting on the rubber elements in the axial direction after assembling the roller.

The mathematical formulation of the optimization problem has the form: to find a vector $\vec{x}$ that minimizes the objective function

$$Z = \min_{\Gamma} \left[\max_{\Omega}(W) \right], \tag{4}$$

where Ω is the cross section zone of the of the element with a variational boundary Γ.

Constraints on variational parameters (geometric) are set by the system of inequations:

$$x_j^- \le x_j \le x_j^+, \quad (j=1, 2,\ldots,m), \tag{5}$$

where x_j^- , x_j^+ are the limits of variation of the components of the vector $\vec{x}$; m – the number of independent parameters.

Functional constraints are set by the system of inequations:

$$\begin{gathered} F_z - [F_z] > 0; \\ \max_{\Omega}\left(|\tau_{rz}|\right) - [\tau_{rz}] < 0; \\ K_{r\min} \le K_r \le K_{r\max}; \\ K_{Z\min} \le K_Z \le K_{Z\max}, \end{gathered} \tag{6}$$

where F_z - the axial force providing the necessary friction force; τ_{rz} - the tangential stresses due to assembly; $K_{r\min}$, $K_{r\max}$ - the lower and upper limits of radial stiffness; $K_{Z\min}$, $K_{Z\max}$ - the lower and upper limits of axial stiffness.

Thus, in the problems of optimal design of the internal damping rubber elements of the roller wheels of caterpillar systems, the objective function and functional constraints imposed on the variational parameters are not set analytically, but in the form of an algorithm [13]. To solve such problems, optimization methods related to direct search methods are used [14-16]. In this paper, to solve optimization problems, the complex Box method is used [14, 15]. This approach is a modification of the Nelder – Mead simplex method [16], which allows one to take into account the constraints on various parameters, both explicitly and functional limitations.

In Figure 5 the finite element model of the cross section of the rubber element of the internal depreciation of the track roller is presented. Tthe structural parameters of that model were obtained as a result of solving the optimal design problem (4-6) as applied to the structure shown in Figure 1. The coordinates of the nodal points of the finite element model located on the inner and outer diameters of the rubber ring were variable. As a result of solving the optimal design problem, the maximum value of the specific strain energy caused by secondary loading by radial force is reduced by 1.49 times compared with the initial design (Figures 4b and 5c). In this case, the tangential stresses caused by the assembly decreased by 1.24 times (Figures 4a and 5b).

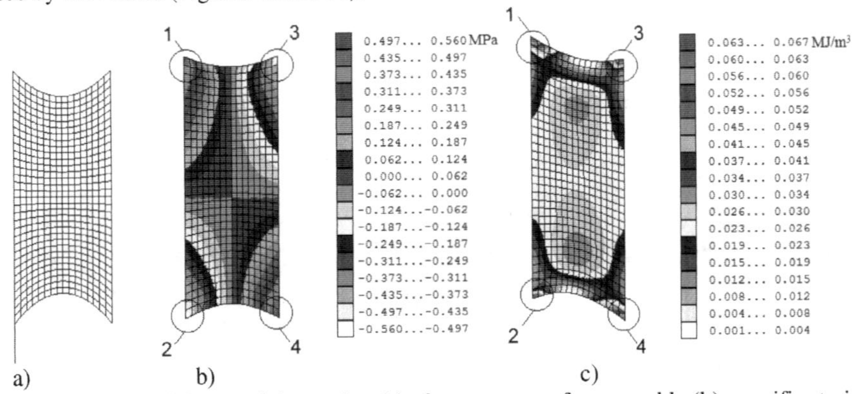

Figure 5. The initial state of the section (a), shear stresses after assembly (b), specific strain energy (c), caused by secondary loading of the radial force

During the search for structural parameters of the section, the finite-element model of which is presented in Figure 6, the coordinates of the nodes located on the end surface of the rubber ring were additionally taken as variable parameters of the optimal design problem (4--6), which allowed us to reduce the maximum value of the specific strain energy caused by secondary loading by radial force compared to the original design by 1.85 times (Figure 4b and 6c). In this case, the tangential stresses caused by the assembly decreased by 1.35 times (Figure 4a and 6b).

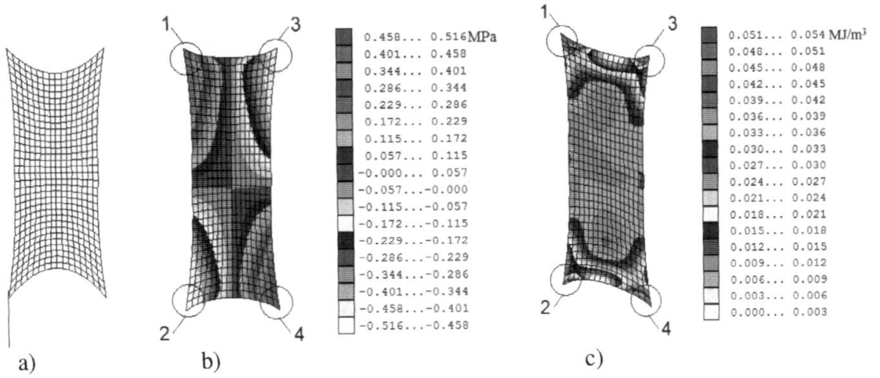

Figure 6. The initial state of the cross section (a), the tangential stress after assembly τ_{rz} (b), the specific strain energy (c), caused by the secondary loading of the radial force

6. Conclusions

As a result of solving the optimal design problem, the shape and design parameters of the internal damping rubber element of the roller wheel of the tracked system were calculated, which made it possible to reduce the specific strain energy under radial loading by 1.85 times. In this case, the shear stresses caused by the assembly decreased by 1.35 times.

IOP Conf. Series: Materials Science and Engineering **695** (2019) 012005 doi:10.1088/1757-899X/695/1/012005

References

[1] Platonov V F 1973 *Dinamika i nadezhnost' gusenichnogo dvizhitelya [Dynamics and reliability of tracked propeller]* (Moscow. Mashinostroenie) p 232

[2] Sharipov V M 2009 *Konstruirovanie i raschet traktorov [Design and calculations of tractors]* (Moscow. Mashinostroenie) p 752

[3] Campanelli M Shabana A A Choi J H 1998 Chain Vibration and Dynamic Stress in Three-Dimensional Multibody Tracked Vehicles *Multibody System Dynamics* **2**(3): 277-316

[4] Choi J H, Lee H C and Shabana A A 1998 Spatial Dynamics of Multibody Tracked Vehicles Part I: Spatial Equations of Motion *Vehicle System Dynamics* **29**(1):27-49.

[5] Lee H C, Choi J H and Shabana A A 1998 Spatial Dynamics of Multibody Tracked Vehicles Part II: Contact Forces and Simulation Results *Vehicle System Dynamics* **29**(2):113-137

[6] Korostelev S A, Verbilov A F and Kovalev V V 2013 Influence of elastic characteristics of the road wheels of tracked system on their dynamic loading *Izvestia of the Samara Research Center of the Russian Academy of Sciences* **15**(2):515-518

[7] Richter E E, Berezin I J 2009 Energy fracture criterion to estimate the fatigue resistance of constructions with elastoplastic *Vestnik Yuzhno-Ural'skogo Gosudarstvennogo Universiteta. Seriya Mashinostroenie* (11):73-78

[8] Hromov M K 1983 On pattern of variations of the in fatigue endurance of rubber *Kauchuk I Rezina* (6):29-38

[9] Zhang J, Xue F, Wang Y, Zhang X, Han S 2018 Strain energy-based rubber fatigue life prediction under the influence of temperature *Royal Society Open Science* 5(10):180951

[10] Treloar L R G 1975 *The Physics of Rubber Elasticity*, (Clarendon Press, Oxford) p 310

[11] Green A E, Adkins J E 1960 *Large Elastic Deformations and Non-Linear Continuum Mechanics* (Clarendon Press, Oxford) p 348

[12] Korostelev S A 2015 Definition the stress-strain state of rubber elements of internal shock absorbers of the road wheels of caterpillar tracks *Izvestia of the Samara Research Center of the Russian Academy of Sciences* **17**(4):793-798

[13] Zhao J, Li Q, Shen X 2008 Finite Element Analysis and Structure Optimization for Improving the Fatigue Life of Rubber Mounts *Journal of Macromolecular Science Part A: Pure and Applied Chemistry* (6): 479-484

[14] Bunday B 1984 *Basic Optimization Methods* (London: Edward Arnold , Bedford Square)

[15] Box M J 1965 A New Method of Constrained Optimization and a Comparision with Other Methods *Computer Journal* (8): 42-52

[16] Nelder J A, Mead R 1965 A Simplex Method for Function Minimization *Computer Journal* (7): 308-313

IOP Conf. Series: Materials Science and Engineering **695** (2019) 012006 doi:10.1088/1757-899X/695/1/012006

Autonomous control system prototype for GAZ A65R32

Tiurikov M. I.[1], Malinin N. V.[1], Evdokimova T. S.[1], Sinodkin A. A.[1] and Fedosova L. O.[1]

[1] Nizhny Novgorod State Technical University n.a. R.E. Alekseev, Nizhny Novgorod, Russian Federation

E-mail: stdbreaks@gmail.com

Abstract
The article presents current state of development of autonomous control system prototype for GAZ A65R32. The solutions for mapping and localization, planning, used sensors, as well as the applied methods of computer vision and deep learning and the methods used to transfer data in the system are described.
Keywords: driverless vehicles, ADAS, computer vision, machine learning, mapping, localization, deep learning

1. Introduction
ADAS systems in an actively developing area that gathers the attention of many researchers and corporations. The main problem of current automatic driving systems is the high cost of equipment, mainly due to the use of lidars to obtain information about the environment. In addition, the development of such systems is carried out sequentially according to the levels of driver assistance systems ADAS, so at the moment such control systems represent the further development of adaptive cruise control systems. The main goal of development described in the article is to reduce cost of such systems by replacing lidars by much cheaper video cameras and stereo cameras and to develop level 5 ADAS system from scratch bypassing intermediate levels.

2. Hardware equipment
We equip out prototype with two stereocameras and lidar to provide mapping module with information about objects and landscape around the vehicle. At this stage of development our sensor configuration does not provide full 360 degree field of view for obstacle perception, as of lack of bandwidth provided by on-board computer of the vehicle.

As the workaround for this problem we gain 180 degree field of view provided by Hokuyo UTM-30LX-EW 2D lidar with working range of 30 meters on the front paired with ZED Camera to get additonal 3D information about obstacles and ground surface at distances up to 20 meters and 90 degrees field of view without giving up too much precision.

Additionaly we have placed Intel RealSense D415 camera at the back of the vehicle to get additional 63 degree rear field of view but with small range of up to 10 meters as on larger distances result point cloud becomes too noisy to provide significant information for mapping module.

To gather data about vehicle's movement and position the UBlox C94-M8P GPS/GLONASS module is utilized along with Atmel XPlained Pro BNO055 development board featuring Bosch BNO055 inertial measurement unit.

As on-board computer for our vehicle we use a regular PC with following configuration:
CPU: AMD Ryzen 5 2400G
RAM: 16 GB DDR4
GPU: 2 x Nvidia GeForce GTX 1060 6 GB
SSD: 250 GB SATA

3. Control system modules implementation
3.1 Mapping module
As mentioned earlier mapping module uses two stereo cameras and lidar to gather information about vehicle's surroundings.

Mapping module consists of three levels of maps as shown in Figure 1:

1. 3D mapping algorithm RTABMap[1] is utilized to create precise 3D map of mostly static objects. As we observed during first tests of mapping system this algorithm is not suitable for highly dynamic environments as it often struggles to fastly cleras freed areas of map.

2. 2D global grid map was added to avoid getting leftover parts from moving obstacles as described above by fuzing lidar and using higher refresh rates than RTABMap. It is used by global planner to provide vehicle with route from current position to goal state.

3. 2D local grid map is built by getting small area of global grid map. It also has higher resolution and refresh rate and relies heavely on lidar data as it provides much higher refresh rates than can't be anchieved by stereocameras. Is is utilized by local planner that is used to avoid obstacles and provide low level control system with steering and speed commands.

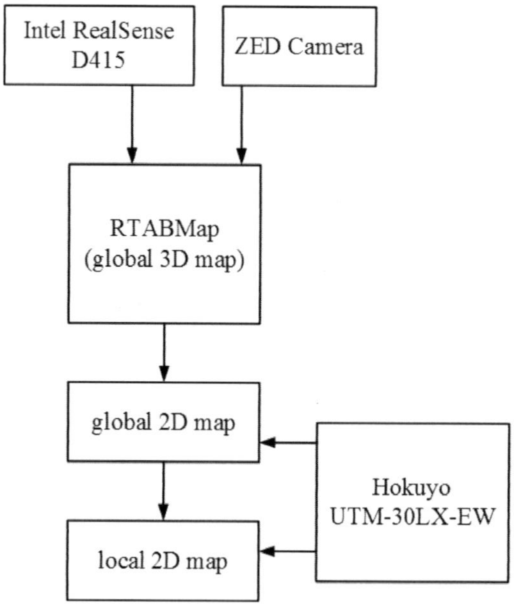

Figure 1. Structure of mapping module.

3.2 Localization module

Localization module heavely relies on visual odometry received from ZED Camera driver. Frame processing is done on the graphics cards by proprietary algorithms provided by ZED with the driver. The optimal resolution of 3840 by 1080 pixels at 30 frames per second is selected based reseach presented in [2].

Additionaly to increase robustness of localization we utilize Atmel XPlained BNO055 module to achieve high refresh rates of vehicle's position, direction, velocity and acceleration. To get the global coordinates the UBlox C94-M8P GPS/GLONASS module is installed so we can set target points in global coordinates and track vehicle's position. The data fusion is made by applying extended Kalman filter.

As a future work we are planning to use Maplab[3] to get visual odometry from cameras as currently used proprietary algorithms from ZED Camera driver requires CUDA capatible GPU to run.

3.3 Planning module

The planning module utilizes the scheme that is common for mobile robotics which constists of global planner to get complete route from current position to goal so local planner do not get stuck in local minima and the local planner to perform obstacle avoidance. The global planner is base on common A*[4] path planning algorithm and the local planner uses Dynamic Window Approach (DWA) [5]. Those algorithms are both simple and use models that do not perfectly describes kinematics of a car, but their simplisity allows us to achieve low latency reaction of the vehicle to dynamic obstacles.

3.3 Machine vision module
As of now machine vision module includes parking lot recognition, signs and traffic lights detection and lane segmentation solutions.

Sigh and traffic light detection along with lane segmentation are performed by convolutional neural networks. To train those networks we mostly utilize BDD100K[6] dataset for detection and segmentation tasks paired with Russian Traffic Sign Dataset (RTSD) [7] for traffic sign classification.

3.3.1 Traffic sign and traffic light detection.
Object detection is a computer vision and image processing task related to semantic localization of objects of a certain classes. This submodule implements detection of objects of two classes: traffic sign and traffic light.

As a metric of accuracy of different detection neural network architectures Average Precision or mAP is utilized:

$$AP = \sum_1^n (r_{n+1} - r_n) p_{interp}(r_{n+1}) \quad (1)$$

where p is precision:

$$Precision = \frac{TP}{TP+FP} \quad (2)$$

and r is recall:

$$Recall = \frac{TP}{TP+FN} \quad (3)$$

here TP is true positive detections count, FP is false positive detections count, FN is missed of false negative detections count.

During architecture search for detection module we considered only single shot detectors such as SSD[8], YOLO[9] and RetinaNet[10]. Most development time was spend on this stage. RetinaNet was selected as a base for out detection module giving 58% mAP score compared to significantly smaller 40% mAP for SSD and 45% mAP for YOLO. The training time for RetinaNet in out case was around 30 hours on a single GTX 1080ti.

The results are present on Figure 2 below.

Figure 2. Combined results of sign detection and recognition module and lane segmentation module.

3.3.2 Lane segmentation.
Semantic segmentation is a process of division of every pixel in an image into a specific classes of objects. Semantic segmentation is used to highlight different objects in an image.

The deep learning is utilized for this task as of classical computer vision algorithms such as WaterShed or MeanShift is not capable of providing such level of accuracy as deep learning approaches.

To compute precision of semantic segmentation neural networks we used Intersection over Union metric or IoU:

$$IoU = \frac{S_O}{S_U} \quad (4)$$

where S_O is area of intersection between true area of object on image and neural network predicted area for this object and S_U is union of true area of object and neural network predicted area of image for specific object.

As in object detection task multiple architectures of neural networks for semantic segmentation were reviewed such as: PSPNet[11], ENet[12] and LinkNet[13]. LinkNet were selected as a base for this module giving us 81% IoU with about 9 times higher frame rates than PSPNet with 89% IoU. To achieve 81% IoU we trained LinkNet for around 25 hours on a single GTX 1080ti.

3.3.3 Parking lot detection.
To detect parking lots for automated parking as of now we utilize classical computer vision algorithms as there is no suitable datasets developed yet for this task to train neural networks.

The overall pipeline consists of preforming Bird's View transform on image from camera, color thresholding followed by detection of vertical and horizontal lines with angel tolerance of 10 degrees, then rectangle is fitted to cover parking lot area and parking lot keypoints are calculated relative to camera position.

The results are present on Figure 3 below.

Figure 3. Parking lot detection results visualization.

3.4 System components integration

Integration of discribed modules is done via Robot Operating System (ROS). ROS includes libraries to exchange messages between modules using TCP sockets, which is implemented as Publisher-Subscriber scheme with support for themes for message topics and some basic message types to ease further development of control systems for robots.

At present moment message exchange system implemented in ROS does not support data encription and client authentication. So to develop a basis for future secure message exchange system of autonomous vehicle we implemented a NATS based message system to exchange messages between high level control system (local planner in particular), low level control system driver and smartphone used for teleoperation. NATS is a high performance message broker which supports authentication, message encryption and transfer over TLS protocol and is supported on wide wariety of platforms. Simplified data flow scheme is present of Figure 4.

Integration of data from machine vision modules is now in active development. Current mapping module does not support any convinient methods to integrate data other than obstacle positions. Our solution of this problem is to reimplement mapping module using multilayered maps provided by Grid Map[14] library. This will allow us to use a dedicated layers to map results of different vision tasks with obstacle information and further consider it in cost functions used by global and local planners. Figure 5 illustrates integration of different sensors to global multilayered 2d maps.

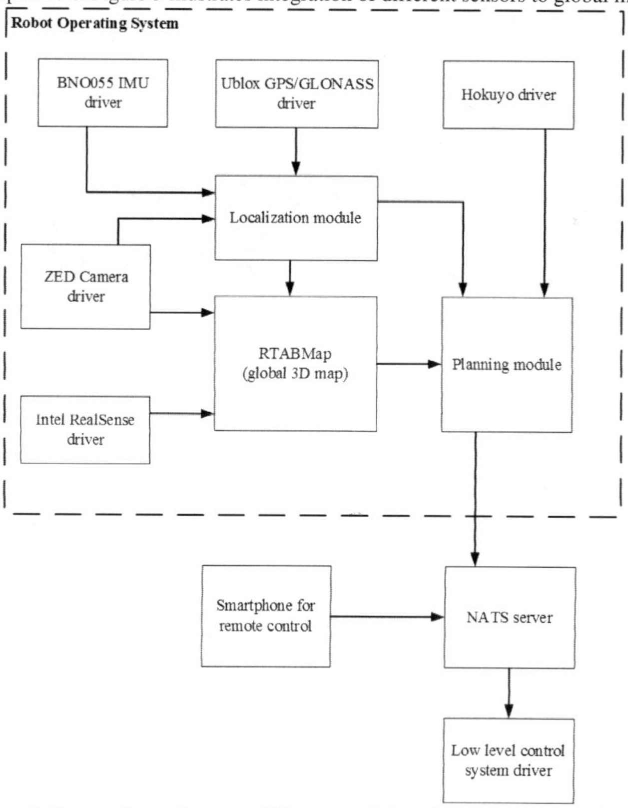

Figure 4. Data exchange between different modules of autonomous driving system

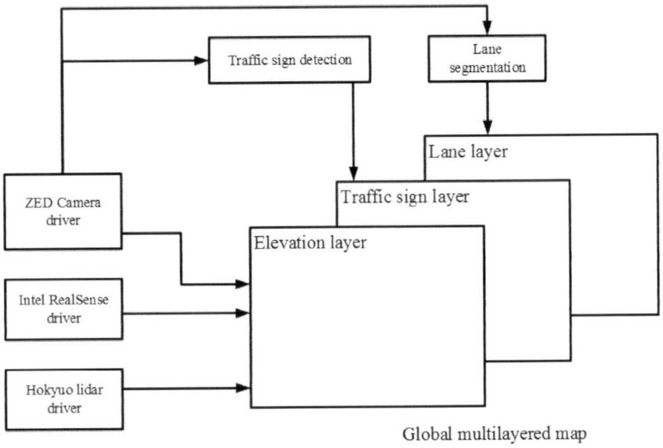

Figure 5. Sensor data integration scheme for multilayered grid maps.

4. Conclusion and future work

Proposed autonomous vehicle prototype was tested on "ROBOCROSS-2019" open test of unmanned vehicles and aircrafts where it took a third place. At this stage our prototype is able to autonomously avoid obstacles and most of future works is related to integration of vision modules data into mapping and planning modules.

As one of future research areas we are targeted at development of highly integrated hardware vision modules based on neural network processors. As part of this research we are planning development of efficient convolutional neural networks for image processing on embedded hardware.

Acknowledgement

This research done with the financial support from Ministry of Science and Higher Education of the Russian Federation (the unique project identifier is RFMEFI57717X0268).

References

[1] Labbé M., Michaud F. RTAB-Map as an Open-Source Lidar and Visual SLAM Library for Large-Scale and Long-Term Online Operation 2019 *Journal of Field Robotics.* **36(2)**

[2] Luis E.O., Elizabeth V.C., Luiz M.G.Depth Data Error Modeling of the ZED 3D Vision Sensor from Stereolabs 2018 *Electronic Letters on Computer Vision and Image Analysis.* **17(1)**

[3] Schneider T., Dymczyk M.T., Fehr M., Egger K., Lynen S., Gilitschenski I., Siegwart R. maplab: An Open Framework for Research in Visual-inertial Mapping and Localization 2018 *IEEE Robotics and Automation Letters*

[4] Hart P.E., Nilsson N.J., Raphael B. A Formal Basis for the Heuristic Determination of Minimum Cost Paths 1968 *IEEE Transactions on Systems Science and Cybernetics* **4(2)**

[5] Fox D., Burgard W., Thrun S. The dynamic window approach to collision avoidance 1997 *IEEE Robotics & Automation Magazine* **4(1)**

[6] Yu, F., W. Xian, Y. Chen, F. Liu, M. Liao, V. Madhavan, and T. Darrell BDD100K: A Diverse Driving Video Database with Scalable Annotation Tooling 2018 *arXiv e-prints, arXiv:1805.04687.*

[7] Shakhuro V.I. Russian traffic sign images database 2016 *Computer optics* **40**

[8] Liu, W., D. Anguelov, D. Erhan, C. Szegedy, S. Reed, C.-Y. Fu, and A. C. Berg SSD: Single Shot MultiBox Detector 2015 *arXiv e-prints. arXiv:1512.02325*

[9] Redmon, J., S. Divvala, R. Girshick, and A. Farhadi You Only Look Once: Unified, Real-Time Object Detection 2015 *arXiv e-prints. arXiv:1506.02640.*

[10] Lin, T.-Y., P. Goyal, R. Girshick, K. He, and P. Dollár Focal Loss for Dense Object Detection 2017 *arXiv e-prints. arXiv:1708.02002*

[11] Zhao, H., J. Shi, X. Qi, X. Wang, and J. Jia Pyramid Scene Parsing Network 2016 *arXiv e-prints. arXiv:1612.01105.*

[12] Paszke, A., A. Chaurasia, S. Kim, and E. Culurciello ENet: A Deep Neural Network Architecture for Real-Time Semantic Segmentation 2016 *arXiv e-prints. arXiv:1606.02147.*

[13] Chaurasia, A. and E. Culurciello LinkNet: Exploiting Encoder Representations for Efficient Semantic Segmentation *arXiv e-prints. arXiv:1707.03718*

[14] Fankhauser P., Hutter M. A Universal Grid Map Library: Implementation and Use Case for Rough Terrain Navigation 2016 *Robot Operating System (ROS) – The Complete Reference*

Calculation and experimental research of liquid flow through a radiator of the electric car power battery thermal managment system

Meshkov A.M., Barinov A.A., Sizov A.Y., Skrebneva T.G. and Korotunova I.A.

Nizhny Novgorod State Technical University n.a. R.E. Alekseev, Nizhny Novgorod, Russian Federation

E-mail: aleksandr.sizov@nntu.ru

Abstract
The article presents current state of the development of the thermostatic control system for electric vehicle traction batteries. The results of calculation and experimental research of the fluid flow through the radiator of the electric car power battery thermal managment system are described. Based on the data obtained, a pressure-flow characteristic of the radiator is constructed.
Keywords: CFD, coefficient of hydraulic resistance, liquid cooling system, electric vehicle battery

1. Introduction

Currently, in the automotive industry there is a steady trend towards the development and implementation of innovative technologies. A major area of development and research in the automotive industry is the transition to environmentally friendly energy sources, instead of the traditionally used carbon-based internal combustion engines. Recently, vehicles with an electric type of engine (electric vehicles) have become very popular. Structurally, the electric car is driven by one or more electric motors powered by an autonomous energy source, which is usually an electric battery. In addition, the battery system, in addition to batteries, also includes charge balancing systems, a battery management system and a battery temperature control system.

The thermal managment system of the battery is designed to maintain acceptable temperature on the battery cells (battery cells, Figure 1) under various climatic conditions of car operation.

Figure 1. Battery cell

The presence of this system is primarily due to the fact that the battery is the most expensive component of the electric vehicle, and the battery cells are the most expensive component of the battery (Figure 2). It is also worth noting that the battery cells can not be repaired, so when they fail, they are replaced with new ones. The service life of the battery cells can be increased by maintaining a constant cell temperature in the range from 20 °C to 25 °C (for most types of battery cells).

IOP Conf. Series: Materials Science and Engineering **695** (2019) 012007 doi:10.1088/1757-899X/695/1/012007

Figure 2. The ratio of the battery cost to the car cost

Experimental research [1] show that at a temperature of 45 °C for 1500 charge-discharge cycles, the ability of a cell to hold a charge decreases by more than 30%, and the increase in internal resistance reaches 100%. However, while maintaining the temperature of the battery cell at 24 °C for 1500 charge-discharge cycles, the ability of the cell to hold a charge decreases by less than 15%, and the growth of internal resistance decreases to 30% (Figure 3).

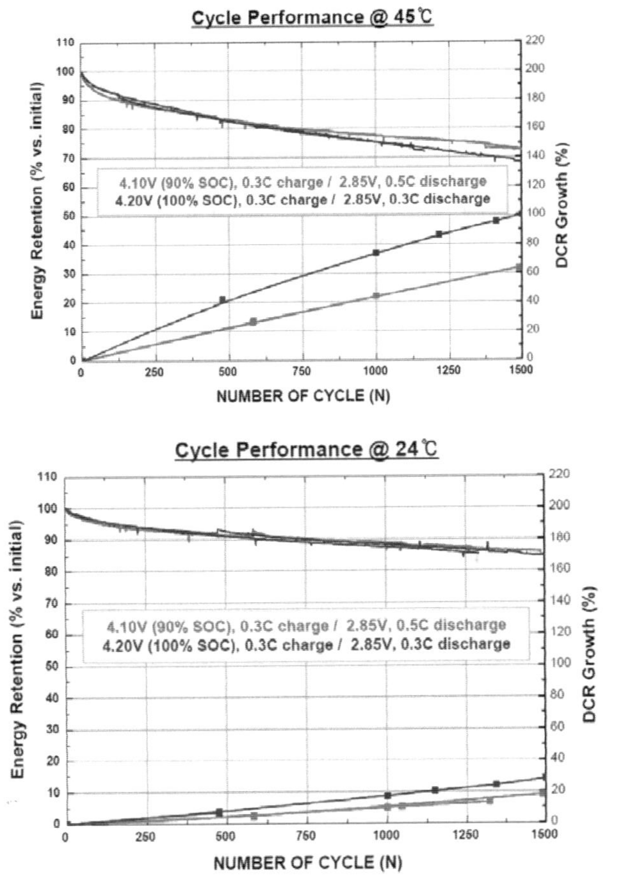

Figure 3. The graph of the dependence of the wear of the battery cell on the number of charge-discharge cycles at a temperature of 45 °C and 24 °C

Taking into account the above factors, it can be concluded that the presence of a battery thermal management system increases the life of the battery cells and reduces the cost of repair and maintenance of the electric vehicle's traction battery.

2. Research subject

One of the main elements of the battery thermal management system is a liquid radiator (Figure 4), which serves to remove or supply thermal energy to the battery cells of the traction battery. These radiators are connected by a single pipe with a circulating coolant.

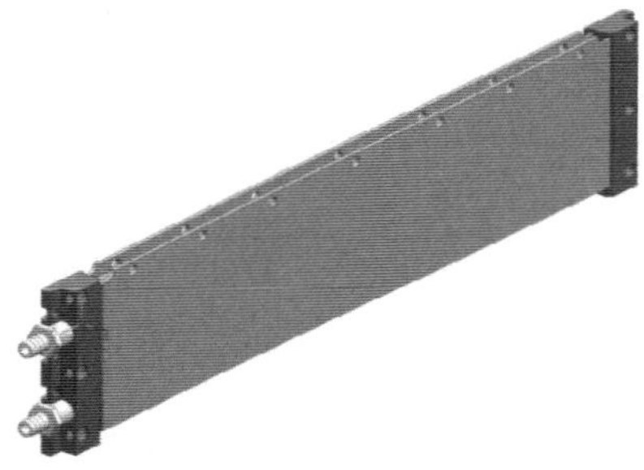

Figure 4. General view of the radiator

One of the main tasks in designing the pipeline of a thermal managment system is to determine the hydraulic resistance coefficient (HRC) of the system (1). This value is necessary to select the operating point for the volumetric flow rate of the liquid in the system and select the necessary circulation pump. One of the main sources of hydraulic resistance in the pipeline of the thermal managment system is the liquid radiator of the battery. Since the battery, depending on the configuration, consists of 10-20 radiators, it is necessary to evaluate the hydraulic resistance of this element with high accuracy. To determine the hydraulic resistance, methods of computational fluid dynamics (CFD-Computational Fluid Dynamics) can be used, which allow modeling the turbulent fluid flow generated inside the radiator channel with fairly high accuracy.

$$\xi = \frac{2 \cdot \Delta P}{\rho \cdot v^2} \tag{1}$$

where ξ is the hydraulic resistance coefficient (HRC), ΔP is the pressure drop (Pa) created by the fluid flowing through the radiator, ρ is the fluid density (kg/m^3), v is the fluid flow velocity in the calculated section (m/s) (in as the calculated one, the section of the pipeline supplying medium is taken).

3. Experimental Research

To confirm the calculation results using CFD analysis, an experimental research was conducted, the purpose of which was to determine the HRC of a liquid radiator.

Figure 5 shows the hydraulic circuit of the bench for hydraulic testing of the radiator. Liquid (water) enters the test radiator 6 from the pressure tank 1 by means of a circulation pump 2. Measurement of the flow rate and temperature of the liquid in the supply line takes place using a flow meter 4 and a temperature sensor 3. The pressure drop across the test radiator is measured by a differential pressure gauge 5. Tubing of the test bench (Figure 6) is made of polypropylene pipes reinforced with fiberglass.

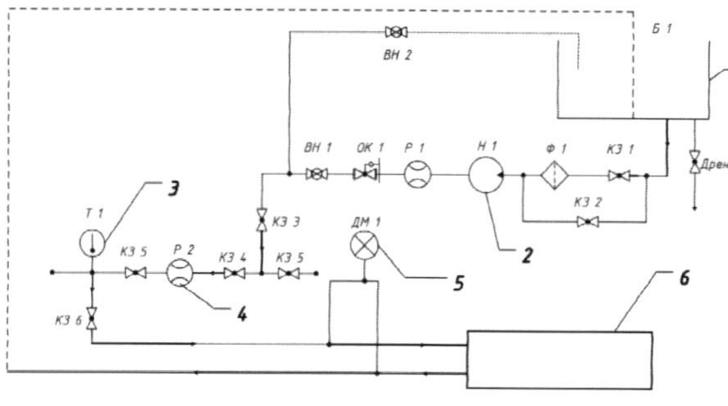

Figure 5. Hydraulic circuit test bench

Figure 6. General view of the test bench

According to the recommendations from [2] and [3], in order to obtain the correct values of the pressure drop (ΔP) generated by the radiator, it is necessary to position the test radiator and the static withdrawals of the differential pressure gauge in areas with a stabilized flow. To fulfill this condition, the withdrawals and the inlet (fitting) of the test radiator are located at a distance L_{start}.

$$L_{start} = 15 \cdot d \quad (2)$$

where d is the inner diameter of the pipe of the hydraulic circuit.

The definition of HRC radiator is carried out in two stages. Since withdrawals are located at some distance from the test radiator, sections of the supply pipes create additional hydraulic resistance. Based on the fact that the purpose of the tests is to determine the hydraulic resistance of the radiator (ξ_{rad}), it is necessary to measure the pressure drop created by the friction of the flow against the pipe walls ΔP_{pipe}) and subtract the calculated hydraulic resistance coefficient (ξ_{rad}) from the total coefficient of hydraulic resistance of the radiator and pipes sections (ξ_{total}).

$$\xi_{rad} = \xi_{total} - \xi_{pipe} \quad (3)$$

$$\frac{2 \cdot \Delta P_{rad}}{\rho \cdot (q/f)^2} = \frac{2 \cdot \Delta P_{total}}{\rho \cdot (q/f)^2} - \frac{2 \cdot \Delta P_{pipe}}{\rho \cdot (q/f)^2} \quad (4)$$

where ΔP_{rad} - the desired pressure drop on the radiator; ΔP_{total} - pressure drop measured with a radiator connected to the circuit; ΔP_{pipe} - pressure drop measured when connecting the supply pipes.

At the first stage, the pressure drop is measured (ΔP_{total}) in the hydraulic circuit system consisting of the tested radiator and supply pipes (Figure 7). The flow meter, temperature sensor and pressure transducer (manometer) are taken.

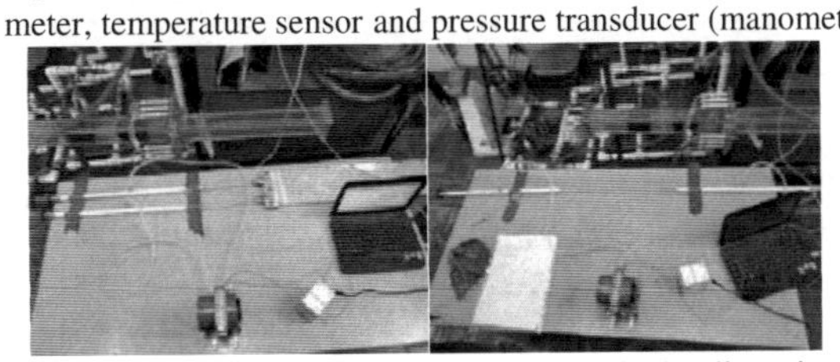

Figure 7. Method for determining the radiator hydraulic resistance

At the second stage, the pressure drop is measured (ΔP_{pipe}) in the hydraulic circuit system without a radiator. For this, the radiator is excluded from the pipeline, and the supply pipes are interconnected, forming a new closed loop (Figure 7). The fluid flow rate is set equal to the flow rate measured in the first test stage. The flow meter, temperature sensor and pressure transducer (manometer) are taken. The measurement results of the volumetric flow rate of the liquid (q, [m^3/s]), pressure drop (ΔP_{pipe} [Pa]), temperature of the liquid in the circuit (T [°C]) are recorded in the table of the test report.

After the experiment, the results are processed and the radiator HRC is calculated (ξ_{rad}). Provided that the results of measuring the pressure drop (ΔP_{total}) and (ΔP_{pipe}) are obtained at the same fluid flow rate (q), we can calculate the value of the hydraulic resistance coefficient of the radiator (ξ_{rad}) based on the ratio:

$$\xi_{rad} = \frac{2\cdot(\Delta P_{total} - \Delta P_{pipe})}{\rho\cdot(q/f)^2} \quad (5)$$

4. Processing experimental data

Table 1 - the results of the calculation of the radiator hydraulic resistance

Where Q is the volumetric flow rate of the liquid (m^3/s), ξ_{rad} is the HRC of the radiator and ΔP_{rad} is the drop in static pressure (Pa) created by the fluid flowing through the radiator.

Table 1 - the results of the calculation of the radiator hydraulic resistance

№	Q, $\cdot 10^{-3}$ (m^3/s)	ΔP_{rad} (Pa)	ξ_{rad}	ξ_{ave}
1	0,169	13041	3,08	3,06
2	0,188	16041	3,08	
3	0,200	18437	3,04	
4	0,224	22578	3,04	

After calculating the hydraulic resistance coefficient of the radiator, it is necessary to determine the measurement errors [4].
Calculation of the systematic measurement errors in determining the HRC of the radiator:

$$\delta\varepsilon_{syst} = \delta\varepsilon_{rad} + \delta\varepsilon_{pipe} \quad (6)$$

Where $\delta\varepsilon_{syst}$ is the systematic error of measurements in the system of supply pipes and the radiator, $\delta\varepsilon_{rad}$ is the systematic error of measurements of the radiator, $\delta\varepsilon_{pipe}$ is the systematic measurement error of the supply pipes.

$$\delta\varepsilon_i = \frac{C}{q_i^2}\cdot\varepsilon_{diff} + \frac{2\cdot C\cdot\Delta P_i}{q_i^3}\cdot\varepsilon_{flow\_met} \quad (7)$$

$$C = \frac{2\cdot f^2}{\rho}$$

where ε_{diff} is the relative error of the differential pressure gauge, ε_{flow_met} is the relative error of the flowmeter, and ε_i is the systematic measurement error.
The calculation results are presented in table 2

Table 2 - The calculation results

ε_{diff}	ε_{flow_met}	$\delta\varepsilon_{rad}$	$\delta\varepsilon_{pipe}$	$\delta\varepsilon_{syst}$
0.005	0.01	0.124	0.032	0.092

Calculation of random measurement error:
We take the sample size of the measurements N = 4, then:

$$\sigma\varepsilon_{ave} = \sqrt{\frac{\sum(\varepsilon - \varepsilon_{ave})}{N\cdot(N-1)}} = 0.017 \quad (8)$$

Student's coefficient K_{st} = 2.353 for n = 1 degrees of freedom and 95% of the two-way confidence interval.
Random measurement error:

$$\delta\varepsilon_{rand} = K_{st}\cdot\sigma\varepsilon_{ave} = 0{,}039 \quad (9)$$

The total measurement error during testing:

$$\delta\varepsilon_{total} = \sqrt{\delta\varepsilon_{rand}^{\ 2} + \delta\varepsilon_{syst}^{\ 2}} = 0.13 \quad (10)$$

5. Calculation research

For the correct description of the calculated geometry by the finite volume method used in computational fluid dynamics (CFD), a number of operations are necessary [6].

To simulate fluid flow through the radiator channel, the region of the fluid flow is generated from the initial three-dimensional radiator model (Figure 8), then the resulting geometry is edited to simplify the construction of the computational grid. The finished computational grid (Figure 8) consists of finite volumes of various types: the regions suitable for constructing the computational grid using the extrusion algorithm are constructed from triangular prisms, the remaining parts of the geometry are constructed from tetrahedral elements, the wall regions were crushed by prismatic elements. This action is necessary for the correct resolution of the wall layer of the fluid flow, in which a high velocity gradient occurs, caused by energy loss due to overcoming the friction forces against the wall of the radiator channel. The total number of finite elements of the primary mesh is approximately 1.6 million. This size was established based on preliminary tests grid convergence of solutions.

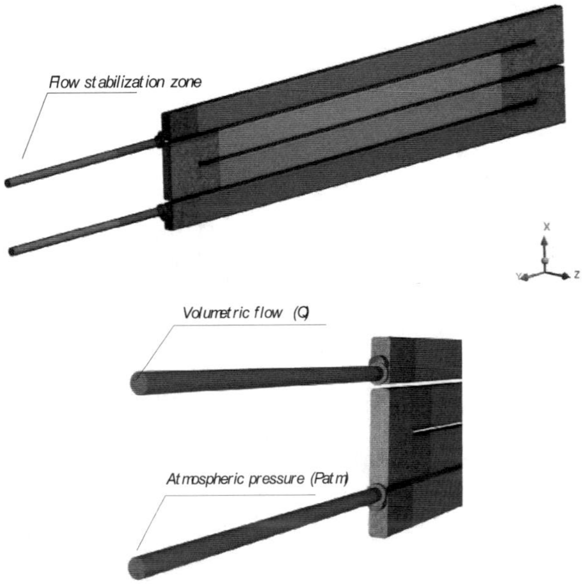

Figure 8. Mesh design model of the radiator

To correctly describe the fluid flow, a two-parameter model of turbulence "k-e Realizable" with standard model parameters was used. The boundary conditions for the calculation are: the volumetric flow rate (q, m^3/s) of the liquid at the inlet section, the atmospheric pressure at the outlet section (R_{atm}, Pa), and the no-slip wall conditions for all remaining external faces. The water under 20°C is selected for fluid domain.

6. Calculation results

According to the data obtained during the calculation research, it is possible to calculate the radiator hydraulic resistance coefficient. For a correct assessment of the pressure drop created by the fluid flowing through the radiator, it is necessary to take the result from the inlet and outlet sections of the radiator, i.e., do not take into account the stabilization sections (Figure 8). The difference in the static pressure values at the inlet and outlet sections of the fluid flow region of the radiator will be the total loss of pressure of the radiator. The calculation results are shown in table 3.

$$\Delta P_{rad} = P_{in} - P_{out} \tag{11}$$

Table 3 - Numerical simulation results

№	Turbulence model	Input speed , v (m/s)	ΔP (Pa)	ξ_{rad}	ξ_{ave}
1	k-e realizable	4,01	25125,5	3,131	3,150
2	k-e realizable	3,6	20421,6	3,158	
3	k-e realizable	3,37	17919,3	3,162	

Figure 6 shows the pressure distribution on the surfaces of the radiator fluid flow region. The greatest hydraulic resistance to the flow is exerted by the zones near the inlet sections formed by the intersection of the fittings and the radiator body. In these areas, the high velocity of the fluid flow relative to the main channel of the radiator, and there is a change in the direction of fluid flow due to the geometric shape of the channel. The distribution of the streamlines of the fluid flow velocity (v, m/s) through the radiator is shown in Figure 7.

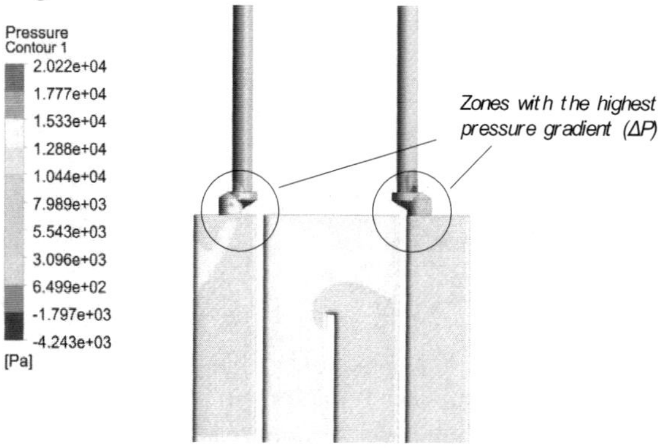

Figure 9. The distribution of static pressure

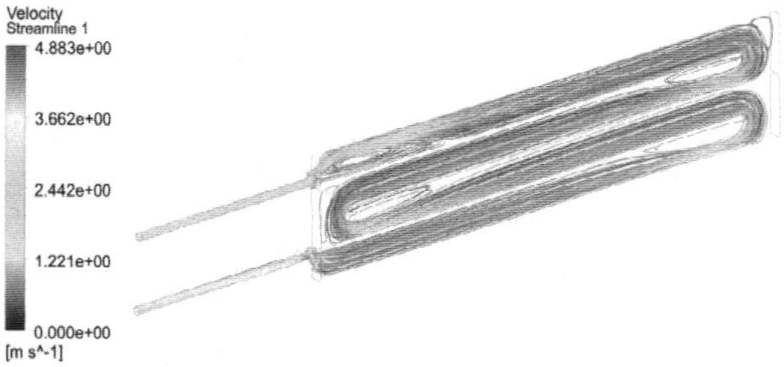

Figure 10. Distribution of streamlines

One of the main hydraulic characteristics of the pipeline elements, in this case the radiator, is the pressure-flow characteristic. The pressure-flow characteristic of the radiator is a graph of the pressure drop (ΔH, m) created by the radiator versus the volumetric flow rate of the liquid (Q, l/s) flowing through the radiator.

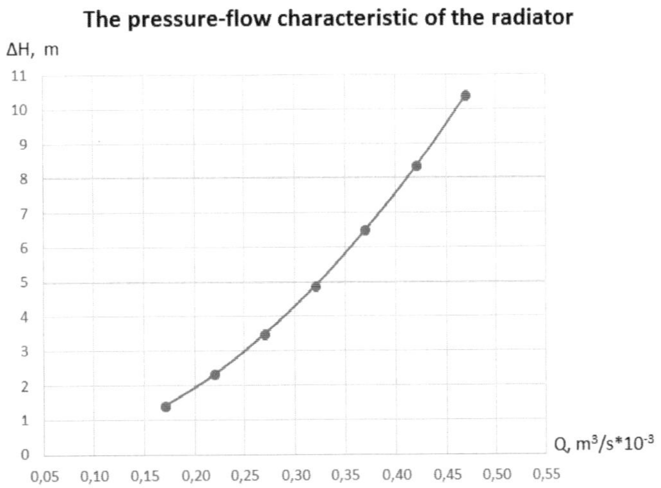

Figure 11. The graph of the pressure-flow characteristics of the radiator

7. Analysis of the of calculation and experimental research results

1. During the experimental research, the value of the radiator hydraulic resistance coefficient $\xi = (3.06 \pm 0.13)$ was obtained; $\delta\xi = 2\%$. This result satisfies the accuracy criteria ($\delta\xi = 5\%$).

2. When comparing the results of calculating the radiator using CFD codes with the results of the experiment, it was possible to find out that the calculation model and the proposed calculation procedure are correct, since $\xi_{ave} = 3.15$ falls within the confidence interval of the experimentally obtained value.

Acknowledgements
This research done with the financial support from Ministry of Science and Higher Education of the Russian Federation (the unique project identifier is RFMEFI57717X0268).

References

[1] Halla F., Touzria J. Experimental in vestigation of the thermal and cycling behavior of a lithium titanate-based lithium-ion pouch cell 2018 *Journal of Energy Storage* **№17**
[2] Idelchik I.E. Handbook of hydraulic resistance / ed. Steiberg M.O. - 3rd ed. add.1992
[3] GOST R55508-2013
[4] Taylor J. Introduction to the theory of errors. Per. From English. 1985
[5] Roache, P.J Verification and Validation in Computational Science and Engineering 1998 Hermosa Publishers.
[6] On the ANSYS customer portal, search for "best practice" [Electronic resource]

IOP Conf. Series: Materials Science and Engineering **695** (2019) 012008 doi:10.1088/1757-899X/695/1/012008

Hardware implementation of the monitoring system and control units of the robotic vehicle "Modular controller"

M.A. Kolobov, D.V. Bushmanov A.V. Okunev A.V. Kocherov

Nizhny Novgorod State Technical University n.a. R.E. Alekseev, 603155 Minina st., 24, Nizhniy Novgorod, Russian Federation

E-mail: powerwolf1995@mail.ru

Abstract

The modular controller is designed for complex automation of vehicles and provides a universal interface for the integration of devices for various purposes. In the basic kit, three modules represent the modular controller: discrete inputs, DC motor control and an ultrasonic sensor controller. All modules have two types of interfaces designed to build a network: CAN (Controller Area Network) and Ethernet. This allows to successfully integrate various equipment. The paper describes the experience of designing modules of a distributed system for operation in an industrial environment. The requirements of industrial standards in the field of electromagnetic compatibility, climatic design and mechanical stability are taken into account.

Keywords: Ultrasonic sensors module, modular controller, DIO, ISO1212, digital input, encoder, DPT driver, switching power supply, planar transformer.

1. Introduction

In the modern world, robotic platforms are widely used both in various industries (transport robots), and for work in hazardous and inaccessible places (areas with chemical and radioactive contamination, underwater research). An important task when building platforms for various applications is the development of a control system. To date, there are two approaches to designing SC. The first is the development of a control system for a specific task, the use of the most rational solutions. The second method is the assembly of the control system from ready-made modules. Both have advantages and disadvantages. Development from scratch allows you to get the most suitable solution for each specific task, but significantly increases the cost of design and commissioning. The second approach, although it is not applicable or redundant in special cases, but reduces the cost of designing a control system, which reduces the complexity and reduces the cost of production. These advantages have allowed the modular control system to become universally used in the design of robotic platforms.

Despite the variety of ready-made modules available on the market, there are tasks for which the offered products either do not fit the reliability criteria or are too redundant in terms of performance or price. That is why it was decided to design our own modules for the control system.

In the framework of this article, we will consider the features of the development of a modular controller using the example of designing a DC motor control module, an ultrasonic sensor module, and a digital input module. The proposed circuit solutions will be considered and the most preferred ones will be selected from the point of view of reliability, module circuits and printed circuit boards will be developed.

The modular controller is a compact, high-performance device for autonomous control of a robotic vehicle. The solution will simplify the design and configuration of unmanned transport systems.

Figure 1 shows the structural diagram of the controller module when connected via the CAN interface.

IOP Conf. Series: Materials Science and Engineering **695** (2019) 012008 doi:10.1088/1757-899X/695/1/012008

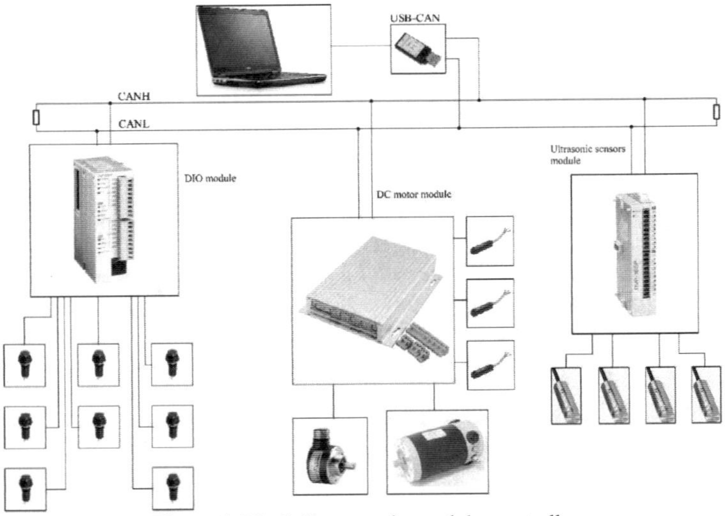

Figure 1. Block diagram of a modular controller

Technical requirements for the modular controller:

1. Supply voltage 12 to 24V
2. CAN, Ethernet interface
3. Waterproof IP rating: IP67
4. Operating conditions +40 to -40 (° C)
5. Galvanic isolation
6. Supply protection, current protection, temperature protection, reverse polarity protection

2. Ultrasonic Sensor Module

Today, ultrasonic obstacle scanning systems are widely used in the automotive industry. However, a common drawback of samples available on the market is the relatively small (≈2.5 m) scanning range, the inability to reprogram both for various modes of polling sensors, and for suitable environmental conditions.

The development of a module of ultrasonic sensors (parking sensors) is due to the need to obtain a reliable and easy-to-use system for determining distances to various objects.

One of the main advantages of the developed module of ultrasonic sensors over existing analogues is:

1. The ability to configure and reprogram the module for a specific task, taking into account the characteristics of the propagation of ultrasound in the environment;
2. Echo processing directly in the sensor itself, which significantly increases noise immunity;
3. Using a transformer to swing the membrane of the ultrasonic sensor, which in turn increases the scanning range;
4. Using the industrial communication interface IO-Link for connecting ultrasonic sensors to the module head board, which significantly increases the distance between the installed sensors.

The main components of the ultrasonic sensor module are shown in Figure 2.

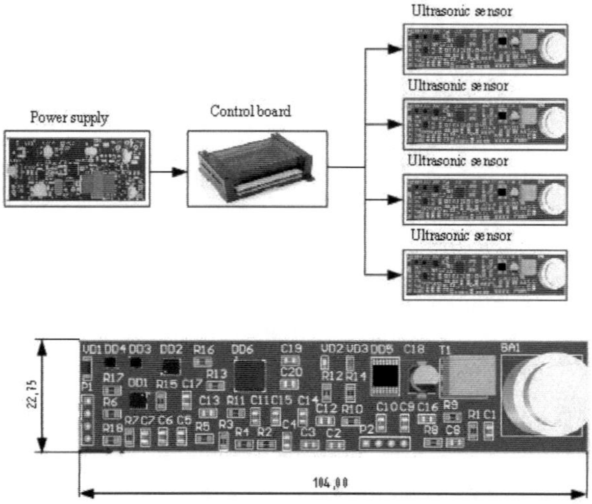

Figure 2. The main components of the ultrasonic sensor module

The propagation of an ultrasonic wave in air depends both on the state of the medium in which the waves propagate and on the parameters of the waves themselves. The state of the environment includes:

1. Air temperature;
2. Humidity;
3. Height above sea level;
4. Flow rate.

The wave parameters include:

1. Ultrasonic wavelength;
2. Ultrasound diffraction;
3. Amplitude of ultrasonic pressure;
4. Reflection from an obstacle.

From all of the above it follows that the state of the environment in which the measurements are carried out should not be neglected, as this greatly affects both the accuracy of the measurement and the possibility of measurement. Also, the received echo signal must be filtered from a lot of noise arising both during the measurement process due to the physical properties of the propagation of ultrasound in the environment, and from noise coming from other objects in the form of unwanted signals. In this regard, it is worth using several types of filters, such as:

1. Band Pass (BP);
2. Peak hold filter;
3. Low Pass (LP) filter.

To achieve this goal, there is a typical solution, which is the Texas Instruments PGA460 family of chips. In accordance with the conditions of the task, a suitable ultrasonic head is selected, and a microcontroller capable of working together with the PGA460 are selected. The microcontroller must read the received data and start the following measurement cycles. Depending on the results of processing the obtained data, the operating parameters of all internal nodes of the PGA460 chip are changed to achieve optimal measurement results.

3. DC motor control module

The DC motor driver is designed for stepless control of the rotation speed of the collector motor over a wide range. The developed module must meet several requirements. Firstly, the developed driver should be easily reconfigured to work with engines of different power in a certain range of values. Secondly, it is necessary to have CAN and Ethernet interfaces, which are common communication protocols for the modules. Since all other control system modules use galvanic isolation between circuit components to increase reliability, it will not be superfluous to increase the reliability of the system in the same way.

The module is planned to be used in robotic platforms, therefore, feedback must be provided. Based on the formulated technical requirements, a structural diagram of the future module was built.

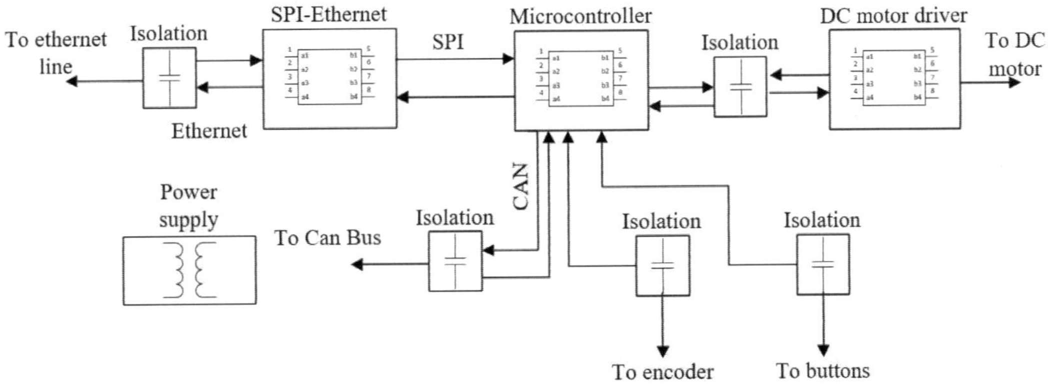

Figure 3. Block diagram of the driver DC

On the diagram, in addition to the microcontroller and, directly, the power key driver, there are CAN and Ethernet modules for receiving control commands, as well as one channel for the encoder and three discrete inputs for feedback. Each of the building blocks is separated from the rest by isolation circuits. Modules should also be powered by a power supply with integrated insulation. We consider each integral module in detail.

3.1 Driver DC

To control DC motors, use the H-bridge circuit shown in Figure 4 (a).

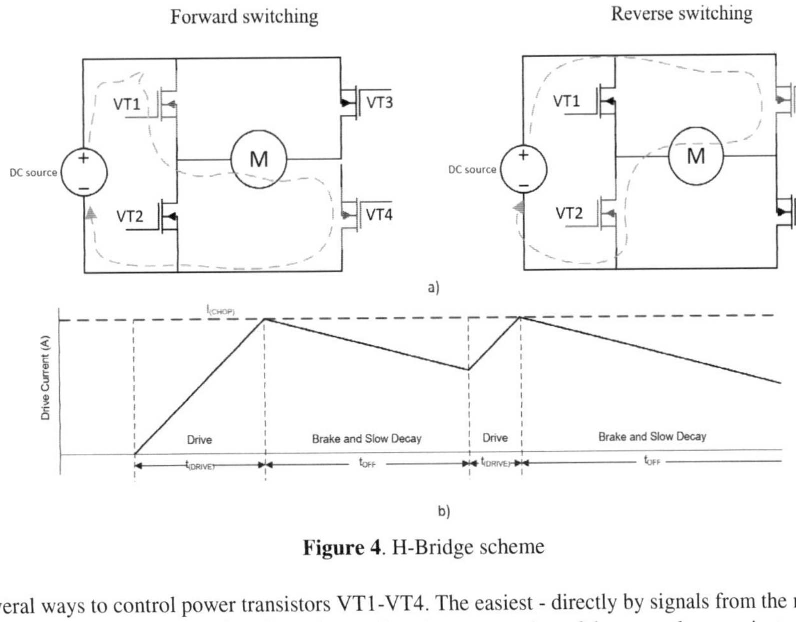

Figure 4. H-Bridge scheme

There are several ways to control power transistors VT1-VT4. The easiest - directly by signals from the microcontroller, or through a simple amplifier. However, with such a scheme, there is no protection of the power keys against overloads, and they can easily fail. An additional circuit that implements current and temperature protection will take up a lot of space on the board. Another control method is the use of integrated circuits, such as DRV8703 from Texas Instrument. A large number of useful systems is implemented inside the chip. The first is the so-called dead time - a small delay during the switching of the keys of the H-bridge in order to avoid short circuits during operation. Secondly, current protection is integrated into the microcircuit, which turns off the power supply of the DCB for a short time, as shown in Figure 4 (b).

When the current through the DCB reaches the value of I_CHOP, a short-term shutdown of the motor power occurs. If, after power is restored, the current again exceeds the maximum permissible value, then the cycle repeats again. The current through the motor is determined by the voltage drop on the current shunt connected to it in series.

The third important advantage of using an integrated solution for power key management is the ability to diagnose errors (overheating of circuit components, excess current through the DCB, various internal errors) via the SPI interface.

SPI-Ethernet Converter

Industrial microcontrollers typically use 2 types of converters with the same characteristics. These are CP2201 from Silicon labs and ENC28J60 from Microchip. Since CP2201 is much more expensive than ENC28J60 (11 cu versus 4.5 cu) and has poorer documentation, we will choose ENC28J60 for use in the driver. The converter allows you to exchange data over Ethernet with a speed of up to 10 Mbps and transmit data in duplex and half duplex mode.

Discrete Input Isolation

For discrete inputs, the data transfer speed is not significant, but when using an encoder, the data flow rate can exceed 1 Mbit / s. Therefore, to isolate the inputs, we use the ISO1212 chip. One chip has 2 digital inputs. It can accept a wide voltage range (9-300V) including AC. The maximum data transfer rate can be 4 Mbps. The chip also has built-in protection against static electricity.

Driver DC Isolation

The power switch control chip operates on the outputs with voltage values of the order of 5V, so the above solution does not suit her. In addition, SPI data rates can go up to 10 Mbps. The ISO7231 chip, which has 2 digital inputs and one output, supporting a data transfer rate of up to 25 Mbps, became the most preferred.

CAN bus isolation

To isolate the CAN bus, a CAN ISO1050 insulator with built-in protection against static electricity and power surges in the data bus itself was used.

Ethernet Isolation

The supply voltage of the ENC28J60 chip does not exceed 3.3 volts. This voltage level cannot damage the microcontroller, therefore, isolating them among themselves is excessive. Danger can come from high frequency interference and static discharges coming through the Ethernet cable. Therefore, in the developed circuit solution, Ethernet connectors with built-in galvanic isolation based on miniature transformers and an RF interference filter were used. Protection against static electricity is provided by a bi-directional protective diode (suppressor) with a breakdown voltage of not more than 6 volts, namely USBLC6-4SC6, designed for use in Ethernet transceivers.

4. Digital input module

The development of a digital input module is necessary for matching the level of various input signals. The following technical requirements were presented to the circuit:

dimensions less than 200x200x100mm;

protection against moisture and dust;

input voltage range 12 V - 60 V;

the module must comply with the standard for electromagnetic compatibility in accordance with GOST IEC 61131-2-2012

To implement the module, three main solutions were considered: a voltage divider, a solution on an optocoupler and using a specialized circuit.

Voltage divider

The simplest scheme is the use of a voltage divider, which has the highest speed. However, the voltage divider is unstable to changes in the input voltage and does not allow to receive signals of different amplitudes equally well. This makes the voltage divider unsuitable for use in level matching circuits used in the digital input modules of industrial controllers. Also, such a solution does not provide galvanic isolation.

Optocoupler solution

Advantage of the solution with optical isolation of components:

- galvanically isolated circuit eliminates the occurrence of spurious currents on the ground bus, which reduces inductive pickups caused by these currents;
- reduction of common-mode voltage at the input of the differential analog signal
- input and output protection against breakdown by large common-mode voltage;

A solution based on an optocoupler has similar disadvantages with a voltage divider, namely: the input stage is a voltage divider in which one of the resistors is replaced by a nonlinear element - a diode. This does not eliminate this decision from the dependence of the response level on the magnitude of the input voltage. This disadvantage can be compensated by installing a current stabilizer in series with the diode. A scheme using an optocoupler provides galvanic isolation, but the speed of data transmission through a potential barrier is determined by the optocoupler.

Solution using a special chip

The most suitable way to solve the problem of digital inputs is to use a specialized chip. The input stage of the microcircuit is a voltage divider, supplemented by a current stabilizer. Galvanic isolation is provided in the following ways: optical isolation of the serial control interface, capacitive or inductive coupling between the input and output stages.

Three chips were considered: CLT01-38SQ7 from STMicroelectronics, ISO1I811T from Siemens and ISO1212 from Texas Instruments.

To provide galvanic isolation in the solution from STMicroelectronics, additional optocouplers ACPL-K73L and ACPL-W70L are required. Siemens does not have a current limit. As a result, an integrated circuit from Texas Instruments ISO1212 was chosen to isolate the digital inputs, as it satisfies the technical requirements for a modular controller.

Figure 5 shows a functional diagram of ISO1212. An external resistor, R_sens, installed at the channel's signal input, precisely sets the current limit and helps minimize system power consumption. Resistor R_THR is used to increase the voltage threshold. The input signal is interpreted as a logic high or low level using a voltage comparator with hysteresis. A hysteresis loop is used to increase noise immunity.

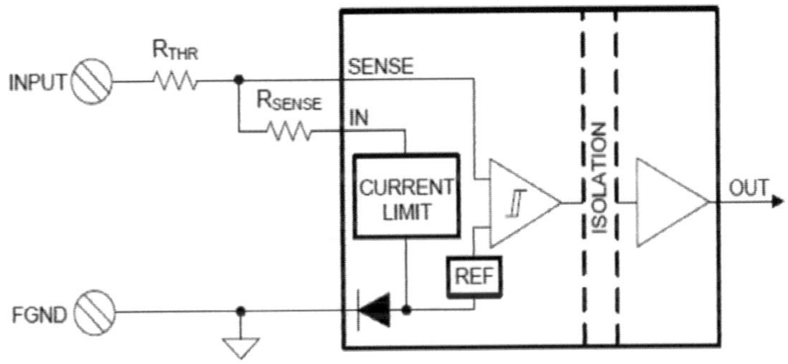

Figure 5. Functional diagram of the ISO1212 chip

The block diagram of the digital inputs and outputs module is shown in Figure 6. In addition to the control microcontroller, the diagram contains CAN and Ethernet modules for receiving control commands from a computer, as well as 8 inputs for connecting buttons and pulse sensors. In order to increase the overall reliability of the circuit, each of the building blocks is separated from the rest using isolation circuits. Modules should also be powered by a power supply with integrated insulation.

To unify the modular controller, the CAN and Ethernet interfaces, as well as the microcontroller and the power supply are used the same as in the DC motor driver module.

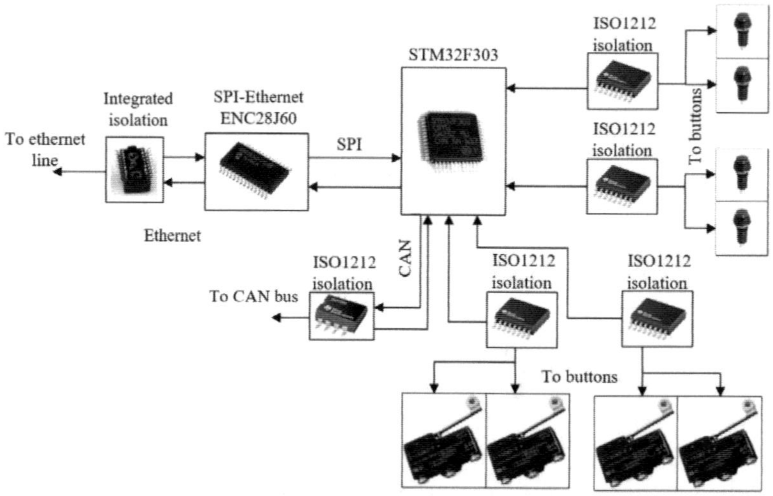

Figure 6. Block diagram of the DIO module

5. Power Supply

Since the modules use data bus isolation, it is also necessary to use separate galvanically isolated power supplies for each structural unit. The control system for each of the modules is constructed in the same way, therefore, it is possible to determine the approximate required power by considering it for only one unit, in this case, for the DPT driver. The data is combined in table 1.

Table 1. The main characteristics of consumers

Name	Number	Maximum current (mA)	Supply voltage (V)
ENC28J60	1	250	3,1-3,6
STM32F303VCT6	1	100	2-3,6
ISO1050DUBR	1	3,6 73	3,3 5
ISO7231CDWR	2	11 27	3,3 5
ISO1212DBQR	3	2	3,3
Encoder	1	100	25 ±5%

It is convenient to develop not separate power supply modules for each structural unit, but one universal power supply unit with separate isolated outputs. We use a transformer as galvanic isolation, which, in order to reduce the dimensions, is planar. For a switching power supply, we use a ready-made circuit based on the UCC28C43D with a sampling frequency of 500 kHz. The calculated power of the pulse converter was about 7 watts, but a certain margin was made on the power components of the circuit, which made it possible to increase the output power. This may come in handy in the future development of more energy-intensive modules.

PCB layout was constructed in Altium Designer. The layout was made in such a way as to minimize the length of the tracks along which high-frequency currents go [5]. The final model of the power supply with its overall dimensions is shown in Figure 7.

Figure 7. General view and overall dimensions of the power supply

6. Conclusion

A module controller design based on the STM32 microprocessor has been developed. Communication with the electronic components of the vehicle is via CAN or Ethernet interface. The following modules are in the process of debugging and testing:

- The module of ultrasonic sensors, which serves to determine obstacles and orientation in a limited space.
- Digital input module that allows you to measure the signal of pulse sensors.
- The module "DC Motor Drivers", designed to control and receive data from the vehicle electric motor.

A power supply module was also designed to power all three circuits.

Acknowledgements
This research done with the financial support from Ministry of Science and Higher Education of the Russian Federation (the unique project identifier is RFMEFI57717X0268).

References

[1] Maniktala S. Switching power supplies from A to Z. St. Petersburg: "Crown Century", 2014-250 p.
[2] Tietze U., Schenk, K. Semiconductor circuitry. 12th ed. Volume I: TRANS. with Deutsche. - Moscow: DMK Press, 2008. - 832 p.: Il.
[3] Tietze U., Schenk, K. Semiconductor circuitry. 12th ed. Volume II: TRANS. with Deutsche. - Moscow: DMK Press, 2008. - 832 p.: Il.
[4] Horowitz P., Hill W. the Art of electronics TRANS. Ed. 2-E. - M.: Publishing house BINOM. - 2014. - 704 p., Il.
[5] Tyutyukov S. Practical recommendations for the development of printed circuit boards
[6] Dementyev Y. N., Chernyshev A. Y., Chernyshev I. A. Electric drive: textbook. - Tomsk: TPU publishing House, 2010. - 232 p.
[7] Zubarev S. PLC: discrete inputs / outputs. Electronics news №1/2019
[8] Nosov Y. R., Sidorov A. S. Optocouplers and their application
[9] Popkov O. Z. Fundamentals of transformative technology: textbook for universities/ O. Z. Popkov. 3rd ed., stereotype. - M.: publishing house MEI, 2010. - 200 p.: Il
[10] Gusev, V. G. Electronics: Textbook for universities-2nd ed., Rev. and add. / V. Gusev, Y. M. Gusev. - Moscow: Higher school, 1991. - 622 p.
[11] Zabrodin, Y. S. Industrial electronics: Textbook for universities/ Yu.S. Zabrodin. - M.: High school, 1982. - 496 p.

Development and implementation of a software and hardware complex of the controls automation into the self-driving car system based on the "GAZelle NEXT" vehicle

M M Voronin, P A Anisimov, A V Zolotov

Alekseev Nizhniy Novgorod State Technical University, 603155 Minina st., 24, Nizhniy Novgorod, Russian Federation

e-mail: mindroz_prosperous@list.ru

Abstract. This article describes the stages and the results of automation of the car controls, as part of the self-driving car based o n the light commercial vehicle (LCV) car class. Some aspects of automation of the car equipped with a manual transmission are touched upon. A block diagram of the automation system of the vehicle controls is provided. The variant of integration of modules of controls automation in the base car is shown. The description of the lower level program is given.

1. Introduction

On the base of NNSTU n.a. R. E. Alekseev and Private Institution of Further Vocational Education «Corporate University "GAZ Group" was formed a project team to create a self-driving vehicle based on LCV-class car "Gazelle NEXT".

Figure 1. Self-driving "Gazelle NEXT"

The first step in creating an unmanned vehicle based on car “Gazelle NEXT” is the automation of its controls - the steering system, gearbox, and pedals, particularly.

One of the most important tasks in creating a self-driving car is to ensure road safety, and this can’t be done without the reliable and fast functioning of all vehicle components. The safety of the designing control automation system was emphasized in its design.

2. General structure description

The proposed structure of the control automation system has a hierarchical structure (see Fig. 2), and it is possible to distinguish the upper and lower levels in general. The *lower level* is represented by direct current motor and stepper motor controllers equipped with sensors. The *upper level* is represented by a computer with a movement planner program and a program that processes the movement planner commands, implementing the control algorithm of the lower level.

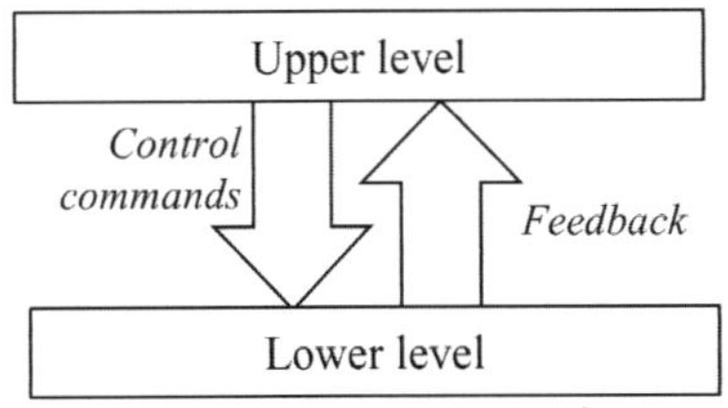

Figure 2. Generalized structure of contro automation system

In the early stages of the design, it was decided to use a personal computer for upper level implementation. The components of the lower level were car controls’ drivers – the controllers of the pedal unit, gearbox, and electric power steering. To provide a communication channel between the upper and lower levels, it was decided to use a CAN interface along with USB-CAN Converter, which was used as a CAN transceiver for the PC. CAN-packets generated at the upper level in response to the movement planner's instructions and the feedback are considered as *control commands* (see Fig. 1). Following that, CAN-packets like module status messages, telemetry packets and sensor readings are considered as *feedback* from the lower level and can be transferred only by request from the upper level.

After specifying the elements of the structure, the control automation system takes the following form (see Fig. 3):

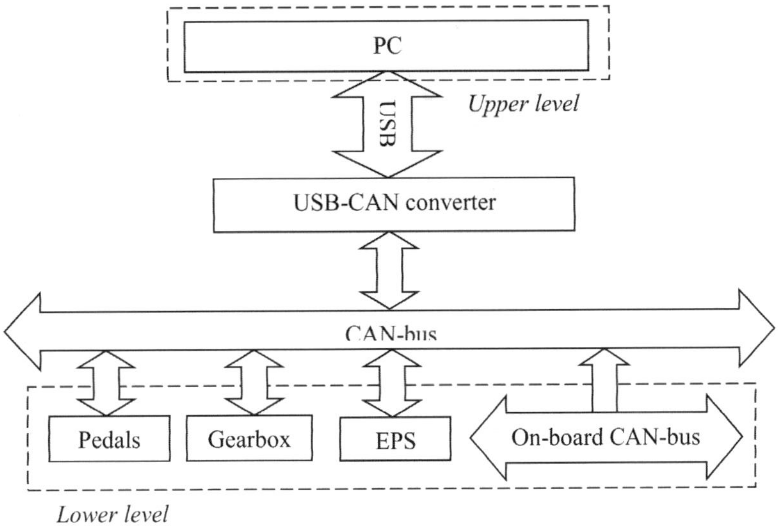

Figure 3. Controls automation system structure

On-Board CAN-bus is used as a source of telemetry of the car: the speed of the machine (in km/h) and engine rate.

IOP Conf. Series: Materials Science and Engineering **695** (2019) 012009 doi:10.1088/1757-899X/695/1/012009

3. Description of the lower level automation units

3.1. Steering automation unit.

To automate the steering in the steering column of the base vehicle was introduced electric power steering «Avtoelectronica» (further EPS). EPS is equipped with a built-in control unit with CAN-interface, as well as direct current motor and absolute steering angle sensors. EPS can operate in two modes: in the manual mode, in which a steering wheel can be controlled directly by the driver, and the engine of EPS performs an auxiliary function, and remote control mode, in which steering is carried out by the control unit that receives commands from the CAN-network.

EPS safety in the remote mode is carried out in the following way: the command source is required to transmit the control synchronization package to the EPS control unit cyclically with a period of not more than 1 second, otherwise, the EPS will go to the manual mode and will inform about the loss of control. Also, the steering wheel goes into manual mode on command from the upper level in case of a critical situation All this was taken into account when writing the algorithm of the main control program of the lower level.

3.2. Pedal block

One way to automate the pedal assembly is to control the pedals with an electric motor converting the rotational motion of the pulley into pedal motion. A stepper motor was used to control the pedals. To ensure the safety of the future system during emergencies, an Autonomous program was created for each of the automation units. For this purpose, the network controller based on the STM32F103 microcontroller was designed. The absolute encoder and stepper motor driver are connected to the controller board. The stepper motor is controlled by CAN commands. The main objective of the can-controlled motor driver controller is to ensure a clear and uninterrupted operation of the self-driving car in the presence of a control signal from the upper level and to perform an emergency stop of the vehicle in case of a system failure.

Figure 4 shows a 3D model of the network controller board, assembled with the case.

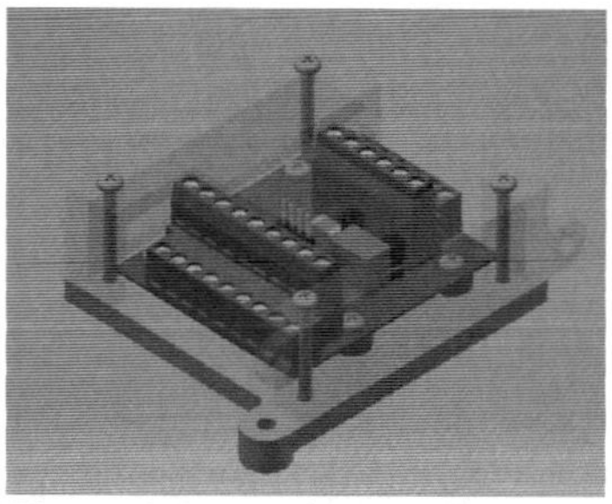

Figure 4. Network controller board

3.3. Gearbox automation unit

The car "GAZelle NEXT" is equipped with a manual transmission. A gearbox automation module is an automated unit that replaces the gearshift knob. The most reasonable and technologically correct solution in this direction is using an automatic or robotic gearbox. However, such a solution for the basic vehicle is not possible, because the car "GAZelle NEXT" is not equipped with such transmission. In this regard, as a module of automation of the gearbox, it was reasonable to develop a unit that simulates gear shifting using the gearbox handle.

Two linear actuators of the TGA-100 model equipped with DC motors were used as an actuating element. The actuators were controlled by CAN-interface using a network controller. This network controller is responsible for timely gear switching on the command from the CAN-bus. The current gear is determined by the limit switches connected to the controller.

IOP Conf. Series: Materials Science and Engineering **695** (2019) 012009 doi:10.1088/1757-899X/695/1/012009

4. Constructive implementation of control automation system modules

4.1. Electric power steering mounting design

To ensure the transmission of torque from the EPS to the steering shaft, the mechanical connection of the EPS output shaft with the standard steering shaft and the steering wheel of the basic vehicle was performed. As a tool for mounting the module into the steering column of the base vehicle, adapter brackets for mounting the module were designed and manufactured (shown in black in Fig. 5, a).

A three-dimensional model of the steering control automation module is presented in figure 5, a. Integrated into the steering column EPS is shown in figure 5, b.

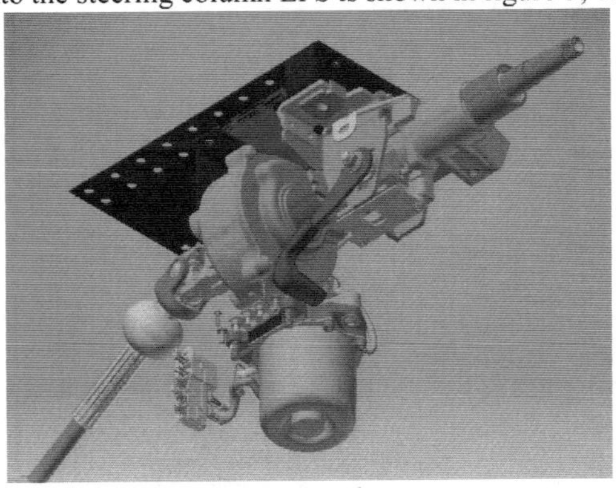

a)

b)

Figure 5. Steering automation unit.

4.2 *Gearbox automation unit design*

To solve the problem of integration of the gearbox automation unit into the control system, the existing gearbox handle was dismantled, and the gear shift was performed directly with the gearbox drive cables. Brackets were designed to secure the cables (see Fig. 6). They were installed on the base support plate of the module - a molded plate of steel sheet with connecting holes. The model of the support plate is shown in figure 7.

Figure 8 shows the model of the assembled gearbox automation unit.

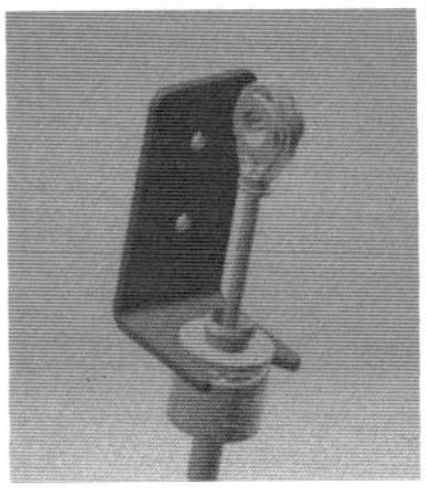

Figure 6. Cabel bracket 3D model

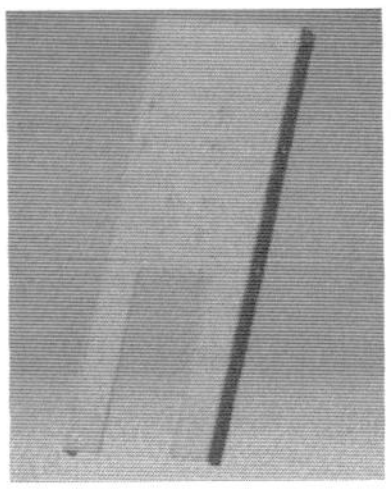

Figure 7. Support plate 3D model

IOP Conf. Series: Materials Science and Engineering **695** (2019) 012009 doi:10.1088/1757-899X/695/1/012009

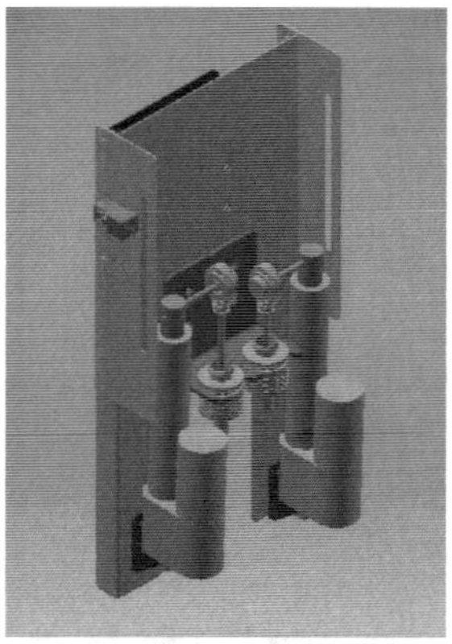

Figure 8. Gearbox automation unit

4.3 Pedal automation unit design

The development of pedal drivers was part of the task of the pedal automation unit integration into the self-driving car control system. The downshift was designed for the implementation of a connection between the stepper motor and the encoder shafts. A single-stage cylindrical gearbox with a gear ratio of 3 was used for this purpose. The model of the reducer with encoder is shown in figure 9. The model of the assembled drive unit is shown in figure 9, b.

a)

b)

Figure 9. Pedal driver unit 3D model

The base plate is installed under the driver's seat is shown in figure 10. A metal cable was used as a kinematic link between the drive and the pedal, which was laid under the floor and the motor shield of the car.

Figure 10. Pedal driver unit inside the car

5. Lower-level control program

The lower-level control program was intended for coordinating the operation of the resulting system modules. The control algorithm assumes the following:

1) Setting up the USB-CAN Converter at program startup;
2) Sending telemetry requests periodically;
3) Processing of received CAN-packets, i.e. telemetry decoding, processing of movement planner commands, error arbitration, etc.
4) Sending command packets to the CAN-bus in real-time;

The program was realized in Python. It is divided into the initialization module and an infinite loop represented by several parallel threads:

1) The main control module, which contains the algorithm of the pedals, gearbox and EPS functioning;
2) Telemetry request thread, which contains the telemetry request sequence;
3) USB-CAN converter dialogue thread, which organizes continuous receiving and sending of CAN-packets;
4) Movement planner commands receiving thread;

In the case of emergency, an emergency message with the highest priority in CAN network is sent, and if it's necessary further stop of the vehicle can be made without the participation of the movement planner, but only with network controllers' possibilities. If the connection between the lower and upper levels is broken, the vehicle stops without the necessary participation of the upper level. Also, the command for switching the EPS to manual mode in case of emergency is sent from the upper level.

6. Conclusion

As a result of this phase of the self-driving car project, the safety of the car automation system was ensured. The reliability of the self-driving car systems was tested in real conditions of operation in the car and showed the stable performance of the tasks. Testing was carried out not only in the conditions of normal system operation but also in case of emergency situations, which indicates the stability of the designed system.

This research done with the financial support from Ministry of Science and Higher Education of the Russian Federation (the unique project identifier is RFMEFI57717X0268).

References

[1] H Asama T Arai T Fukuda T Hasegawa 2002 *Distributed Autonomous Robotic Systems 5* Springer-Verlag Tokyo
[2] D S Vdovin and I O Khrenov 2019 IOP Conf. Ser.: Mater. Sci. Eng. 534 012016
[3] Pfeiffer O Ayre A Keydel C 2008 *Embedded Networking with CAN and CANopen* Copperhill Technologies Corporation
[4] Waschl H Kolmanovsky I Willems F 2019 *Control Strategies for Advanced Driver Assistance Systems and Autonomous Driving Functions* Springer International Publishing
[5] Leandro C Fernandes, Jefferson R Souza, Gustavo Pessin, Patrick Y Shinzato, Daniel Sales, Caio Mendes, Marcos Prado, Rafael Klaser, André Chaves Magalhães, Alberto Hata, Daniel Pigatto, Kalinka Castelo Branco, Valdir Grassi Jr, Fernando S Osorio, Denis F Wolf (2014) *CaRINA Intelligent Robotic Car: Architectural design and applications* Journal of Systems Architecture Volume 60, Issue 4, April 2014, Pages 372-392
[6] Ionita Silviu 2017 IOP Conf. Ser.: Mater. Sci. Eng. 252 012098
[7] Barabás I 2017 IOP Conf. Ser.: Mater. Sci. Eng. 252 012096
[8] Shahian-Jahromi Babak et al 2017 IOP Conf. Ser.: Mater. Sci. Eng. 224 012029

IOP Conf. Series: Materials Science and Engineering **695** (2019) 012010 doi:10.1088/1757-899X/695/1/012010

Prediction of a residual operating life of engines

Y Bazhenov[1] and M Bazhenov[1]

[1] Vladimir State University n a Alexander and Nikolay Stoletovs, 87 Gorky Str., 600000, Vladimir, Russia

E-mail: bagenovyv@mail.ru

Abstract

Results of the executed researches on prediction of an individual residual operating life of engines are pre-sented in article. Prediction is based on the diagnostic information on technical condition of engines ob-tained during the carrying out their maintenance and repair. The analytical equation, the describing regu-larities of change of diagnostic parameters on an operating time and its graphic interpretation are given. The forecasting technique of a residual operating life of mechanisms of engines with use of the lower fidu-cial bound of distribution of their practices to the limiting condition is offered. The technique is approved on the example of forecasting of a residual operating life of cylinder-piston group of the ZMZ-40524.10 engine.

1. Introduction

Prediction of a residual operating life is one of primal problems of diagnosing of a condition of any object including the internal combustion engine (ICE) during maintenance or repair. Definition of a stock of correct work of ICE allows if not completely to avoid opera-tional refusals, then it is essential to reduce their quantity due to timely detection and elimination of the damages which arose in use.

Process of prediction of an individual residual operating life of units and units of the car provides realization of the following stages:

– data collection and processing on operational dependability of the studied object;

– justification of a complex of the diagnostic parameters which are adequately reflect-ing its technical condition;

– development of the analytical equations describing regularities of change of diagnos-tic parameters on an operating time;

– statistical assessment of the predicted residual operating life.

Results and Discussion

The tests executed in actual practice operation on operational dependability of ZMZ-40524.10 engines showed that exhaustion of the operating life put at its development is caused by gradual accumulation of various damages (wear, fatigue and corrosion destructions, residual deformations etc.). Therefore, at prediction of a residual operating life of engine sys-tems change of diagnostic parameter on an operating time with a sufficient degree of accuracy can be described by a power function:

$$S(t) = S_{in} + vt^{\alpha} \quad (1)$$

где S_{in} – initial value of diagnostic parameter; v – intensity of change of diagnostic parame-ter on an operating time t; α – the exponent defining dependence of diagnostic parameter S on operating time t.

Graphic interpretation of this dependence is presented in Figure 1.

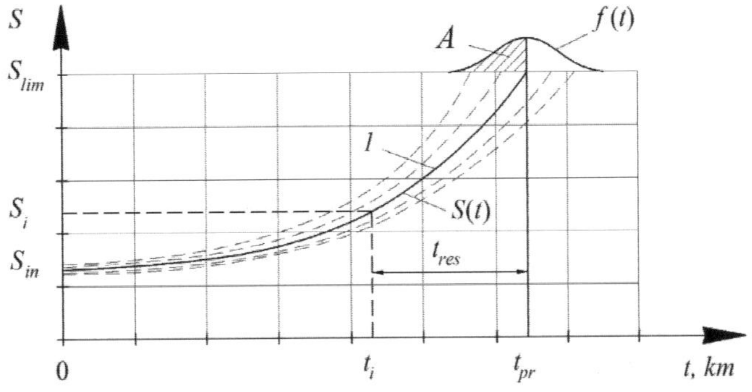

Figure 1. The schedule of change of diagnostic parameter on an operating time: S_{lim}, Si , S_{in} – limit, current and initial values of diagnostic parameter; t_{pr} – predicted operating time to the limiting state; ti – current value of an operating time; t_{res} – residual operating life; f(t) – density function of operating time to the limiting condition

From the provided scheme it is visible that the residual operating life of object t_{res} rep-resents the difference between the predicted value of a time to failure and duration of its op-eration preceding the predicted t_i period.

$$t_{res} = t_{pr} - t_i \tag{2}$$

Thus, for definition of a stock of correct work, it is necessary, by an extrapolation method, according to measurement of diagnostic parameter during the period previous pre-dicted to determine its further change by an operating time to the limiting value and to devel-op the analytical equation approximating this change.

The value of the limited time to failure because of influence of a large number of fac-tors of operation is a random value which can be described by a differential cumulative distri-bution function of *f(t)*. Therefore, actually predicted limiting value of a time to failure repre-sents some average value (on the scheme it corresponds to realization 1), which is defined from expression:

$$t_{pr} = \sqrt[\alpha]{\frac{S_{lim} - S_{in}}{v}} \tag{3}$$

As showed the executed researches, when using as the limit operating time its mean value t_{pr} a stock of correct work at a part of engines it will appear less than calculated (in Fig-ure 1 – area A). It leads to errors of prediction and, as a result, increase in number of engine failures on operation. For decrease of negative consequences of such mistakes in this work it is offered to determine a residual operating life not by mean value of t_{pr}, and by the lower confi-dence limit of distribution of this operating life which is defined from expression:

$$t_{low.lim} = t_{pr} - t(p) \cdot \sigma_t \tag{4}$$

where $t_{low.lim}$ – the lower confidence limit of the predicted time to failure; *t(p)* – Student's fac-tor; σ_t – average standard deviation of operating time.

Student's factor is determined by the given confidence probability which value for the structural elements of the car which are immediately not affecting traffic safety can be accept-ed equal p = 0.80 [1, 3].

Expression (2) for definition of a residual operating life on the lower confidence limit will take a form:

$$t_{res} = t_{low.lim} - t_i \tag{5}$$

Approbation of the developed technique of definition of the predicted residual operat-ing life it is feasible on the example of cylinder-piston group of the engine with use of the di-agnostic S_g parameter (a consumption of crankcase gases) when passing the engineering servic-ing on an operating time of 128 thousand km (at procedural frequency of maintenance the basic car for the III category of an operating conditions t_m = 16 thousand km).

Rated and limiting values of the diagnostic parameters estimating a ZMZ-40524.10 engine cylinder-piston group (CPG) condition are established by manufacturer (Table 1).

Table 1. Standard values of the diagnostic parameters estimating an engine CPG con-dition

No.	Diagnostic parameter	Rated	Limited
1	S_c – pressure at the end of compression stroke, MPa	1.18	0.94
2	S_a – the relative air leakage at position of the piston in UDC within not less than 5 seconds, MPa	decrease from 0.147 to 0.098	decrease from 0.147 to 0.074
3	S_g – consumption of crankcase gases at 4000 min^{-1}, l/min, no more	22	62

By results of control and diagnostic works the following values of the diagnostic pa-rameters estimating technical state of CPG were received:

- pressure at the end of compression stroke, MPa – S_c 1.02
- the relative air leakage, MPa – S_a 0.089
- consumption of crankcase gases, l/min – S_g 29.6

Exponent α and intensity of change of parameter v are determined by practical con-sideration on the basis of processing of statistical material of field tests of change of diagnos-tic parameters by a car operating time. For the S_g parameter of their value made: $\alpha = 1.2$; $v = 0.059$.

The power function describing regularity of change of the S_g parameter on an operating time will take a form:

$S_g(t) = 22 + 0.059 \cdot t1.2$

The predicted value of an average operating time to achievement of CPG of the limit-ing state:

$$t_{pr} = \sqrt[\alpha]{\frac{(S_{lim} - S_i)}{v}} = \sqrt[1.2]{\frac{(62.0 - 29.6)}{0.059}} = 190.9$$ thousand km.

Standard deviation of an operating time to failure of CPG:

$$\sigma_t = \sqrt{\frac{\sum_1^n (t_i - t_{pr})^2}{n-1}} = \sqrt{\frac{6980}{50-1}} = 11.94$$ thousand km,

where n is a sample size of the surveyed engines.

Student's factor at a fiducial probability $p = 0.80$ and number of degree of freedoms of $N = 2$ is equal to $t(p) = 1.29$.

Value of the lower confidence limit of the predicted CPG operating time to failure:

$t_{low.lim} = t_{pr} - t(p) \cdot \sigma_t = 190.9 - 1.29 \cdot 11.94 = 175.5$ thousand km.

Graphic interpretation of definition of the predicted residual operating life of cylinder-piston group is presented by the scheme in Figure 2.

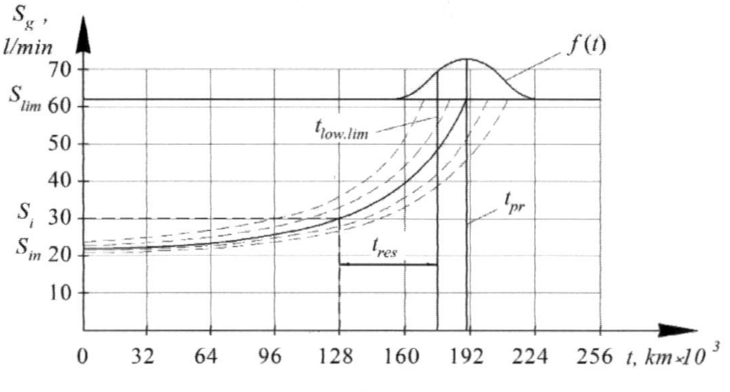

Figure 2. Scheme of definition of a residual operating life cylinder-piston group of the en-gine: t_{pr} – mean predicted CPG operating time to failure; $t_{low.lim}$ – lower confidence limit of the predicted operation time to failure; t_{res} – residual operating life.

The predicted CPG residual operating life on the lower confidence limit will make:

$t_{res} = t_{low.lim} - t_i = 175.5 - 128 = 47.5$ thousand km.

The predicted CPG residual operating life of the studied engine exceeds the estab-lished frequency of an upkeep of the basic car (tm = 16 thousand km) therefore there is no need for profound monitoring of its technical condition and carrying out any technical influ-ences. CPG won't reach the limiting state on the predicted residual operating life $t_{res} = 40.7$ thousand km with the given fiducial probability $p = 0.80$.

IOP Conf. Series: Materials Science and Engineering **695** (2019) 012010 doi:10.1088/1757-899X/695/1/012010

Conclusion

Prediction of a residual operating life on the lower confidence limit of distribution of operating times allows to calculate a stock of correct work of ICE with the given probability, it is essential to reduce its refusals in use and the costs of maintenance connected with it.

References

[1] Bazhenov Y V 2017 Bases of a dependability theory of vehicles *M: FORUM* 320
[2] Bazhenov Y V, Bazhenov M Y 2018 Research of operational dependability of automobile engines *Dependability* **18** No 4 22– 27
[3] Boldin A P 2010 Dependability and technical diagnostics of the rolling stock of the motor transport: manual for higher education institutions *M.: MAII* 206
[4] Bolotin V V 1984 Prediction of a operating life of cars and designs *M: Mechanical engineering* 312
[5] Denisov A S 2007 Ensuring dependability of autotractor engines *Saratov: Publishing house SGTU* 422
[6] Kuznetsov A S 2011 Maintenance and diagnostics of an internal combustion engine *M: Publishing house Academy* 80
[7] Kuzmin N A 2011 Technical operation of cars: regularities of change of serviceability *M: FORUM* 208
[8] Lukinskiy V I 1991 Prediction of dependability of cars *Polytechnics* 222
[9] RD 50-690-89 1990 Dependability in technique. Methods of assessment of indexes of reliability according to the experimental datas *M: Publishing house of standards* 22
[10] Sidnyaev N I 2011 Theory of scheduling of an experiment and analysis of statistical data *M: Urait Publishing House* 399

IOP Conf. Series: Materials Science and Engineering **695** (2019) 012011 doi:10.1088/1757-899X/695/1/012011

Assessment of the technical condition of the units of mobile robotechnical complexes based on the approaches of nonlinear dynamics and artificial intelligence

Yu.G. Kabaldin, D. A. Shatagin, D. A. Sidorenkov, M.V. Zhelonkin, and M. S. Anosov

Alekseev Nizhny Novgorod State Technical University, Nizhny Novgorod

E-mail: uru.40@mail.ru

Abstract
Neural network algorithms for monitoring and diagnosing the technical condition of parts of the MRC drive axle are considered. The formalized task is a comprehensive assessment of the technical condition in the neural network basis. An engineering technique is proposed that can be applied at the stages of bench tests of critical components and assemblies.
Keywords: Technical condition; monitoring and diagnostics; learning algorithm; neural networks

In recent years, robotics is undergoing a new stage of development and covers more and more applications in both the civilian and military spheres, as well as in special working conditions. For this purpose, research is underway to develop mobile robotic systems (MRC) for various purposes. This requires the solution of new fundamental research problems related to improving the intelligence and reliability of MRC. Both the competitiveness of domestic developments in the international market and, in general, Russia's prospects in the context of scientific and technological progress depend on the success of solving these tasks.

Modern MRCs have a large number of embedded diagnostic systems. To identify irregularities in the operation of MRC, in particular, as a result of the failure of individual components or critical parts that can lead to its breakdown during movement, the diagnostic process allows you to detect malfunctions in it and in the future avoid expensive repairs. Therefore, the use of diagnostic methods and tools allows us to assess the technical condition of the main MRC components and ensure its operation is more efficient and effective.

However, given the complexity of dynamic interactions on the vehicle and in the presence of defects in critical nodes, the use of traditional methods for their diagnosis is insufficient. A number of studies have shown that the main problem in diagnosing technical equipment is the chaotic nature of the processes of wear and destruction of parts, assemblies and assemblies during operation. However, such irregular chaotic loads in the nodes of the vehicle cannot be effectively identified by traditional methods and the level of uncertainty of the technical state of the system during operation is quite high. [1-3].

A promising approach to the development of diagnostic systems for critical MRC nodes is the use of artificial intelligence, nonlinear dynamics and fractal analysis approaches. Nonlinear dynamics approaches make it possible to quantify the nature of the motion of complex systems such as MRCs and evaluate their stability at high loads. The use of artificial intelligence approaches allows us to establish a relationship between the features of the diagnostic and operational parameters of the vehicle, given the large amount of incoming information in real time, to evaluate the current technical TS of the system and optimize its operation modes.

To test the dynamic system of the MRC drive axle, a diagnostic stand for critical components and assemblies was designed. The stand consists of a PC with software, a speed indicator, a load regulator, an ADC with vibration sensors, a temperature sensor, etc. The signal sampling rate was 30 kHz, and the operating frequency band was 1 ... 15 kHz.

IOP Conf. Series: Materials Science and Engineering **695** (2019) 012011 doi:10.1088/1757-899X/695/1/012011

Figure 1. Test bench

For signal processing, a method for processing the Vibration signal has been developed, which allows one to identify and classify pulses in the Vibration signal using approaches of nonlinear dynamics and artificial intelligence. The software structure is a series of blocks performing various tasks.

For signal processing, a method for processing the Vibration signal has been developed, which allows one to identify and classify pulses in the Vibration signal using approaches of nonlinear dynamics and artificial intelligence. The software structure is a series of blocks performing various tasks.

For signal analysis, the envelope spectrum method was used. The method has high sensitivity, information content and noise immunity. For signal analysis, a digital filter is used that limits the bandwidth of 1-10 kHz.

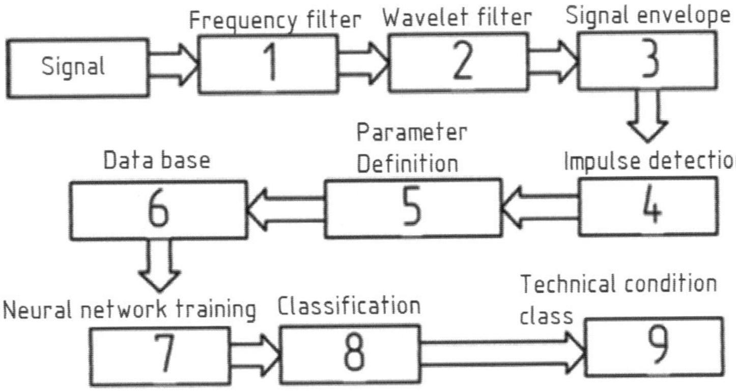

Figure 2. Software functional diagram

After the selection of all signal pulses, their parameters are determined. Parameters such as time maximum amplitude (A), frequency (F), entropy (E) and fractal dimension (D), the highest Lyapunov exponent (λ) are calculated. In addition, an attractor and wavelet scalogram are built for a more detailed analysis. The obtained characteristics of the pulses and the pulses themselves are recorded in a special database of the technical condition of the parts of the MRC unit.

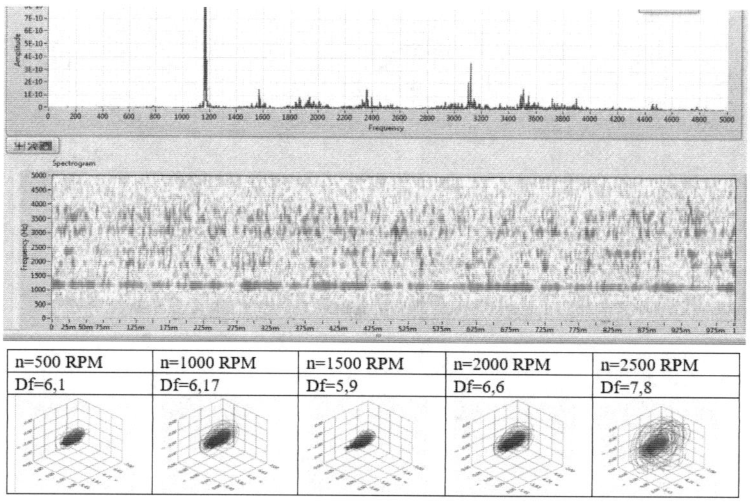

Figure 3. Database Fetch Example

Information from the obtained database is used to train an artificial neural network for the task of clustering. Learning takes place without a teacher using Kohonen self-organizing cards.

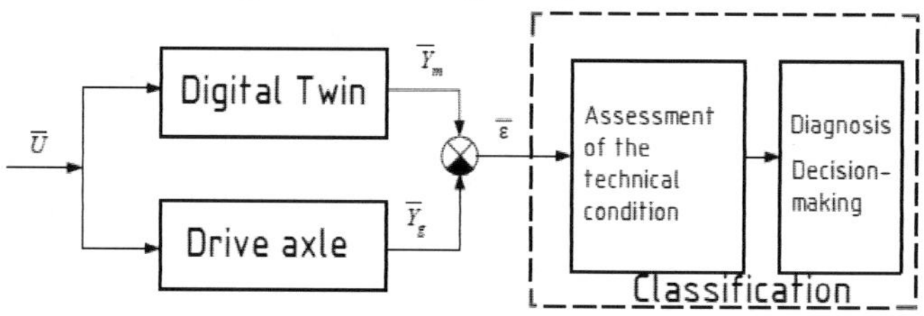

Figure 4. Classification block diagram

The proposed methodology for solving the problem of diagnosing the technical condition of parts of the MRC drive axle is based on a method based on comparing the results of measurements of the Vibration signals of a physical object with the calculated parameters calculated by its mathematical model (digital twin). [6].

In fig. A typical implementation of the method is shown, where U is the vector of control actions; Ym is the vector of parameters calculated using the digital double of the rear axle of the MRC; Yg is the vector of engine output parameters obtained by measuring with sensors; ε = Yg − Ym is the error, which is the difference between the vectors Yg and Y m.

As a mathematical model, a digital twin of the MRC drive axle is used. A digital double is a combination of the use of neural network and numerical simulation using CAD, FEA, FMEA models.

Information on 5 parameters was used as diagnostic features of the input vectors: maximum amplitude (A), frequency (F), entropy (E) and fractal dimension (D), the highest Lyapunov exponent (λ) - X {x1, x2 ... x5}. As the output vectors, diagnostic features of the form D {d1, d2 ... dm} were used, where d is the technical condition expressed in the coefficients of bearings and gears.

To teach the recognition of defects in neural network modeling, five classes of states of the leading bridge were identified: C1 operable state; C2 Inoperative state; C3 Good condition; C4 Malfunctioning condition; C5 Failure.

From the results presented below it can be seen that the overall classification accuracy of the developed neural network classifier is 99.2%. Moreover, the classification accuracy of the 1st class was 89.9%, of the 2nd, 3rd, 4th, 5th classes - 100%.

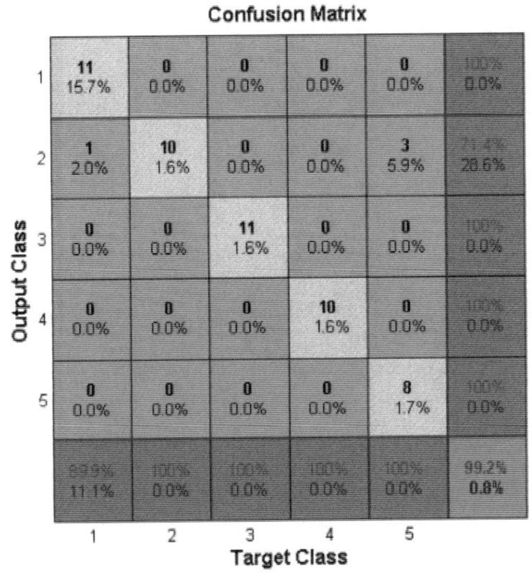

Figure 5 . Classification results

Investigations were carried out both on serviceable parts and on parts with simulated defects. Defects were modeled by assembling a bridge using obviously defective parts. The frequency of rotation of the output shaft is 1200 rpm. Measurements were carried out both when the bridge was operating without load, and under load (braking torque) of 50 N * m.

When the defect reaches its maximum value, the value of information entropy, fractal dimension, and the senior Lyapunov exponent increase. This is explained by the fact that with an increase in the defect the system goes beyond its stable state, the informational entropy of the vibration signals emitted by the system increases, which is typical for chaotic systems. The attractors of the system in this case have the form of partially destroyed limit cycles with obvious chaotic perturbations. The Lyapunov exponent is changed in such a way that the stability of the system is maximum with serviceable parts. As the defect deepens, Lyapunov's senior indicator increases.

The signal entropy of the gear pairs is largely dependent on the load transmitted by the gear pair. At idle, the value of the entropy indicator is negligible. As the forces transmitted by the gearbox increase, entropy increases. This is due to increased vibration from gearing.

This feature of the work of a gear pair to identify trends in state change requires measurements at the same, and large, load. If the load is insufficient defects may not occur. If measurements that differ in time are performed at different loads of the gearbox, then all these measurements will be unsuitable for comparison with each other, for searching for changes that have occurred in the gearbox. In this regard, it is necessary to take into account the load on the node, an indicator of which is informational entropy.

Acknowledgements

An important property of the presented methods is a geometric generalization of the dynamic properties of systems, which facilitates understanding and perception of the features of their dynamic behavior. These geometric generalizations of the dynamic properties of systems are especially relevant at the present time - the time of development of computer technology for automated analysis and synthesis of dynamic systems.

The above methods to improve the assessment of the technical condition of critical MRC units and assemblies will allow for operational monitoring with high performance, as well as more reliable assessment of the current state of the system.

References

[1] Petrov V. A. On the mechanics and kinetics of macro-destruction // Solid State Physics. 1979. - T. 21. - No. 12. - S. 3681 - 3686.
[2] Elchaninov G.S. The methodology for predicting the performance of complexly loaded engineering structures: an abstract of the dissertation of the candidate of technological sciences. St. Petersburg State Polytechnic University, St. Petersburg, 2011.
[3] Artobolevsky I.I., Bobrovnitsky Yu.I., Genkin M.D. Introduction to the acoustic dynamics of machines. - M .: Nauka, 1979.- 296 p.
[4] Golovko, V.A. Neural networks: training, organization, application / V.A. Golovko // Neurocomputers and their application: textbook. allowance / V.A. Golovko. - M., 2001 - 256 s.
[5] Vibrations in technology: Reference: In 6 vols. - M.: Mechanical Engineering, T.1.-1978.-352 p .; T.2.-1979.-351 p .; T.3.-1980.-544 p .; T.4.-1981.-509 s .; T.5.-1981.- 496 p .; T.6.-1981.- 456 s.

IOP Conf. Series: Materials Science and Engineering **695** (2019) 012011 doi:10.1088/1757-899X/695/1/012011

[6] Khaikin, S. Neural networks: full course / S. Khaikin. - M.: Williams, 2006 .-- 1104 p.
[7] Pavlov B.V. Acoustic diagnostics of mechanisms. - M.: Mechanical Engineering, 1971. - 224 p.
[8] Anil, Jacob Diagnosing gears using a new method for monitoring the condition of rotary equipment; translation from English / Jacob Anil, Dr. Y. I. Sharaf-Eldeen. - [Electronic resource].
[9] Yavlensky A.K., Yavlensky K.N. Vibrodiagnostics and prediction of the quality of mechanical systems. - L.: Mechanical Engineering, 1983.- 239 p.

Unmanned Vehicle LIDAR-camera Sensor Fusion Technology

Kuzin V.D., Pronin P.I., Zarubin D.N., Shashkina K.M.

Nizhny Novgorod State Technical University n.a. R.E. Alekseev, 24, Minina St., Nizhny Novgorod, 603950, Russia

E-mail: chromium32@mail.ru

Abstract
With the growth of computing power and the development of information technology in the last few years, the tendency for the development of unmanned vehicles has increased. Many leading corporations sponsor this industry. Since the creation of the concept of an unmanned vehicle implies that the computing systems in its composition must make decisions about the method and direction of its further movement, predict the further actions of the traffic participants and move in accordance with the rules of the road, these vehicles are equipped with powerful telemetry systems that allow to accurately evaluate surroundings. But, since there is a problem in incomplete data from one device (sensor), in this article, using a specific example, sensor fusion technology is considered that allows combining data from several sensors to achieve maximum results in solving problems of developing an unmanned vehicle.
Keywords: LIDAR, camera, sensor fusion, unmanned vehicle

1. Unmanned vehicle

If we consider the telemetry system of an unmanned vehicle, then it can be divided into sections corresponding to a specific task in solving the problem of unmanned traffic. For example, the localization and positioning system and the inertial system make it possible to understand the position in space of an unmanned vehicle.

Groups of optical sensors and radars allow to learn about the parameters of obstacles in the path of movement of an unmanned vehicle and collect data about the surroundings and other participants in the movement. An important factor in the implementation of an unmanned vehicle system is the ability to identify obstacles along the path of an unmanned vehicle using certain sensors or a group of sensors and take actions to prevent a collision. Most systems use obstacle detection algorithms using modular stereo cameras, deep vision systems. In this configuration, an active laser rangefinder of the optical range - LIDAR - is used as an obstacle detection sensor.

The principle of operation is similar to the principle of the radar - a directional signal, and in this case a light laser beam, is emitted by the device, reflected from the object and captured by a highly sensitive sensor. The response time of the signal is directly proportional to the distance to the object in the field of view of the device. Light waves are subject to scattering in any medium, including in the air, therefore it is possible not only to determine the distance to dence (light-reflecting) discrete targets, but also to fix the light scattering intensity in transparent media, which makes it possible to determine in some way such parameters of an object as its illumination, texture, in some cases even color. This article discusses the application of sensor fusion technology to a group of optical sensors, namely, the LIDAR bunch and camera used in the telemetry system of an unmanned vehicle based on the Gazel "Next" (figure 1(a)) and Gazel "Next-drive" (figure 1(b)).

Figure 1. (a) Gazel Next "Vahtan", 1(b) Gazel "Next-drive"

1.2. Optical Sensor Group

In this configuration, LIDARs from Velodyne are used. Testing and implementation of the technology was carried out using VLP-16 models[1]. The standard ROS[2] driver module for working with LIDAR Velodyne VLP-16 (figure 2) publishes data from the device by topic in the ROS framework used in development.

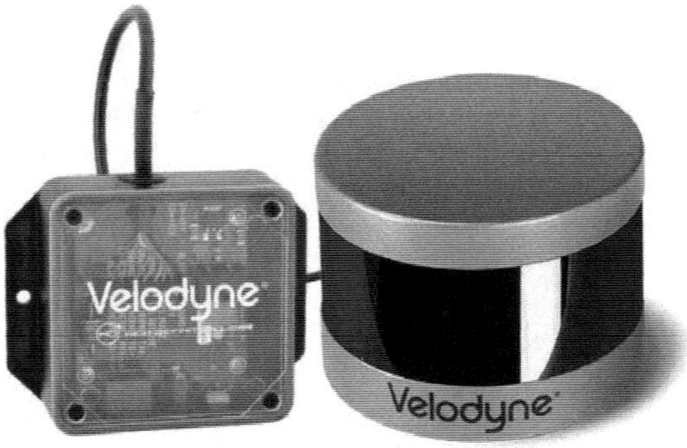

Figure 2. Velodyne VLP-16

The input data is the so-called point cloud, which is a three-dimensional map of the sensor's environment in its field of view. In the current implementation and production of LIDAR, a three-dimensional map with a horizontal viewing angle of 360 ° is built on the hull of an unmanned vehicle. Each point of the cloud contains the necessary and sufficient information for further calculations, namely:

a. Intensity. Intensity - the degree of reflection of the sensor beam from the surface, for example, from a solid light surface, the reflection intensity will be higher than from a black surface.

b. Coordinates in space. Each point has a spatial coordinate in the format (x, y, z) relative to the sensor itself.

c. Belonging to the ring. VLP-16 is a sixteen beam sensor; accordingly, 16 light (laser) emitters rotating in a circle are provided at the physical level of the sensor design.

The Basler[3] acA1300-200uc camera (figure 3) was also used. This camera, equipped with the necessary lens, allows you to receive high-quality video images with a high frame rate, which is important when installing on an unmanned vehicle.

IOP Conf. Series: Materials Science and Engineering **695** (2019) 012012 doi:10.1088/1757-899X/695/1/012012

Figure 3. Basler camera acA1300-200uc

2. Preparatory stages

The sensor fusion task itself involves combining data from multiple sensors. But in the process, it is important to take into account the peculiarity of the operation of these sensors and carry out the necessary preparatory steps. When working with the camera and LIDAR, one of these steps is calibration.

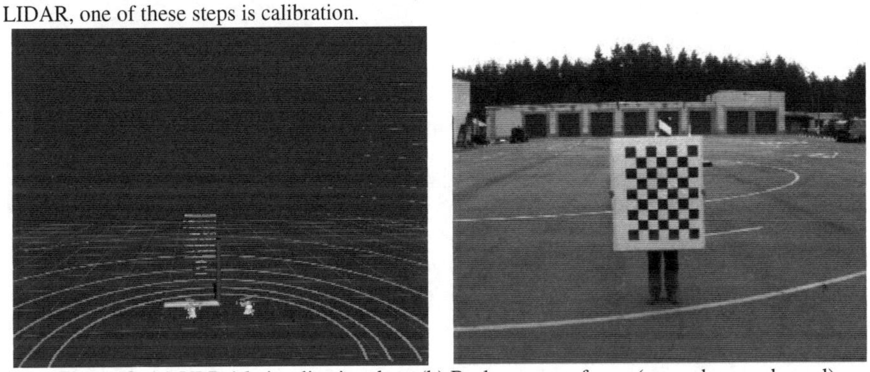

Figure 3. (a) VLP-16 visualization data, (b) Basler camera frame (monochrome channel)

2.1. Camera calibration

Before making measurements in the field of computer vision, it is worth considering the distortion of the image by the camera itself. Calibration (figure 4) corrects complex distortions, which are determined by the shape of the lens, errors in its mechanical fixation, errors in the manufacture of the sensor. Further steps to combine data and image processing should be carried out only after calibration, otherwise calculation errors and associated distortions are inevitable. Calibration of the camera is carried out in order to find out the internal parameters of the camera, as well as the projection matrix used in the further algorithm for joint calibration of LIDAR and the camera.

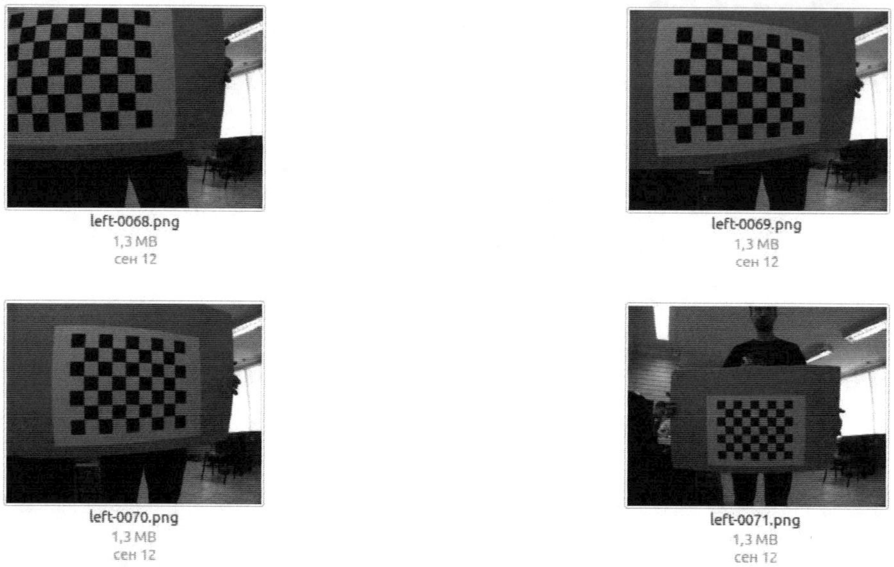

Figure 4. Camera calibration process

After successfully performing camera calibration, having a calibration matrix, you can get a corrected image without the distortion effect.

2.2. Calibration of LIDAR and camera

Calibration of the camera and LIDAR consists in finding the correspondence of their fields of view and the correct setting. In contrast to the known methods of automatic calibration using a perforated panel[4], it was decided to develop another solution to this problem for a more optimal approach. This solution is the development of manual calibration of LIDAR and the camera. This software uses the PCL[5] library and the LIDAR point cloud processing functions it provides for the subsequent transformation of point clouds. Selecting the transformation parameters (yaw, pitch, roll) of the spatial vector of the input cloud and the displacement along the coordinate axes, we can accurately correlate the data of the point cloud on the spatial position of the object in the LIDAR field of view with the position of the object on the camera frame.

3. Sensor fusion

When the sensors are calibrated, the data merge can be performed directly. The combined fields of view of the two sensors allow to expand the range of detection of objects. So, for example, specifically in this case, SF is used to determine the physical parameters of objects in the field of view of these devices with their detection and recognition by the neural network in the image.

Also, the used functionality allows you to directly visualize LIDAR data on the image frame in accordance with the intensity at the current point (figure 5(b)). This approach helps with preparatory steps when it is necessary to find the correspondence of the field of view of two sensors.

How it works. The image from the streaming video from the camera is fed to the input of the neural network. That in turn determines the objects and passes the list of detected objects further to the SF module along with the coordinates of the boundaries of the detected objects on the frame. Further, the module, having received data on the boundaries of objects in a 2D image, draws a correspondence between the object in the 2D field of view of the camera and the 3D view of LIDAR. After completing these steps, the output gives the physical parameters of the desired object directly in 3D - the distance of the object, its size, etc. This approach allows you to work with segments of the LIDAR point cloud as separate objects of space and classify them according to the classes that the neural network provides at the recognition stage. As one of the stages of this work, we can identify the limiting volumes (figure 5(c)) of objects immediately in 3D and use the identified boundaries along the unmanned vehicle's path, for example, to correcate the trajectory of its further movement.

IOP Conf. Series: Materials Science and Engineering **695** (2019) 012012 doi:10.1088/1757-899X/695/1/012012

Figure 5. Steps of sensor fusion:
(a) clear frame, (b) projected neural network data and LIDAR data to the frame, (c) visualized at Rviz limiting volumes of objects

Future work

The immediate prospects are optimization of matching algorithms and matching projections of sensor data, increasing processing speed and improving visualization algorithms. The described sensor fusion methods and algorithms are implemented in the unmanned vehicle control program complex and are currently being tested.

This research done with the financial support from Ministry of Science and Higher Education of the Russian Federation (the unique project identifier is RFMEFI57717X0268).

References

[1] Velodyne VLP-16. Website (https://velodynelidar.com)
[2] ROS – Robotic Operation System. Website (https://www.ros.org)
[3] Basler camera product. Website (https://www.baslerweb.com/en/)
[4] but_velodyne ROS package. Website (http://wiki.ros.org/but_velodyne)
[5] PCL – pointclout library. Website (http://pointclouds.org)

Investigation of collision warning possibilities by means of GNSS receivers of Android smartphones

Roman Kulikov[1], Alexander Chugunov[1] and Vladimir Zamolodchikov[1]

[1] Department of Radio systems, Institution of Radio engineering and Electronics,
National Research University MPEI, Moscow, Russia

E-mail: KulikovRS@mpei.ru

Abstract
Using of raw GNSS measurements in Andoid smartphones for precise mututal positioning of vehicles on a road is considered in the Paper. Field experiments with two moving vehicles and pedestrian were conducted. 15 kinds of general maneuvers were provided (overtaking, turn, oncoming traffic, etc.). Smartphone measurements were logged and post-processed to get mutual positioning estimations during maneuvering. Smartphone estimations were compared with precision GNSS receivers measurements.

Keywords: GNSS, Android, relative positioning, mutual trajectory, collision warning, vehicles, connected cars

Introduction

Road safety and its enhancement is one of the most important topics of Intelligent Transportations Systems paradigm. There are plethora of approaches being investigated and offered. Most of them are based on additional on-board sensors, besides that the most sophisticated are based on inter-vehicle communication and sharing of measurements [1]. Others are based on substational modification of road infrastructure.

Necessity of additional hardware implementaton (sensors, computers, communication equipment) is the main obstacle for new road safety and driver assist technologies scaling. At the same time almost every driver, at least in large cities, already has and uses a smartphone most of which are Androids. Unfortunatelly, general smartphones didn't provide enough accuracy for collision warning purpose.

There were attempts to adjust smartphones for precise posiyioning task, but they required specific hardware and/or low-level firmware modification [2, 3]. Also smartphones could be very suitable for inter-vehicle communication needed for example for RINEX (Reciver INdependent Exchange format) sharing to provide precise GNSS (Global Navigation Satellite Systems, such as GLONASS, GPS, Beidou, Galileo and augmentations like QZSS, SDCM, etc.) positioning in differential modes [4].

In the middle of 2016 Google annonced new generation of Andoid operating system, 7.0 Nougah which allowed access to raw GNSS measurements. That provided potential adaptation of smartphones for application requires precise positioning but deep investigating of smartphones' features regarding to the topic is necessary.

One of the first papers on the subject was [5], and authors demonstrated very promissing results: decimeter-level positioning uncertainty and cm/sec-level velocity uncertainty obtained with smartphone's in-built GNSS receiver. Number of other researches were conducted with different models of Android smartphones [6 – 10] which showed sub-meter acccuracy even without sophisticated filtering as well as main chalenges were identified: poor performance of antenna and battery economy (so called duty cycles).

In [11, 12] an integration of raw GNNS measurements with raw inertial measurements were implemented on smartphones: sub-meter uncertainty and ability of operationg with uncomplete GNSS satellite constellation were demonstrated.

In [13, 14] opportunities of two-band GNSS receiver of smartphone were investigated and PPP (Precise Piont Positioning) mode with cm-level uncertainty was demonstrated as well as reflectometry.

Up to now (August 2019) 53 models smartphones of 13 brands allow access to raw GNSS measurements. 17 of them provide carier phase measurements and 8 of them have two-band GNSS receivers [15]. So, market share of such smartphones grows and one could expect 60..80% share within 2023..2025 at least in most of high-income regions.

There is an opportunity of collision warning by sharing raw GNSS measurements between neighboring driver's smartphones (connected car concept), its processing (sort of RTK, Real-Time Kinematics, with mobile baseline) and prediction of mutual trajectories. For fast scaling this processing is to be performed as plug-in software for leading mobile apps for drivers. However, the above mentioned reseaches mostly were provided in ideal environment (static mode, clear sky) and no one in relative mode between moving objects with sky schielded with metall roof as is is in general use case of smartphone assisting to a driver.

In this paper we present some early results of research we're conducting regarding to usage of smartphones for collision warning of vehicles.

Methodology

Two vehicles (Hyundai Solaris and Audi Q3) were equipped with smartphones (both were Samsung Galaxy S8 with Exynos 8895V chipset) being tested and fixed on driver's panels as it usually done. Both vehicles also were equipped with reference navigation-grade GNSS receivers (NV08C RTK-M) with extermal two-band antennas (Fig. 1).

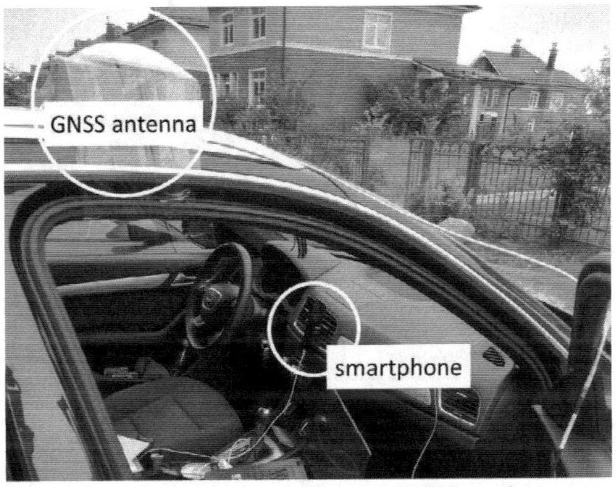

Figure 1. Placement of smartphone and antenna of reference GNSS receiver on one of the testing vehicle.

Besides that, a third reference GNSS receiver was placed static near test road to be binded to absolute ECEF coordinate frame.

Both smartphones and both mobile reference receivers and static reference receiver wrote raw GNSS measurements in RINEX log files. Smartphones logged with help of Geo++ Rinex Logger free mobile app. Referecnce GNSS receivers looged with customized proprietary software.

There was series of experiments with 15 different trajectories of pair of vehicles and 4 different trajectories of vehicle and pedestrian (Fig. 2a, 2b), pedestrian was equipped with the same kit: Samsung Galaxy S8 and Navis NV08C RTK-M with external antenna. Vehicles moved with speed up to 40 km/h.

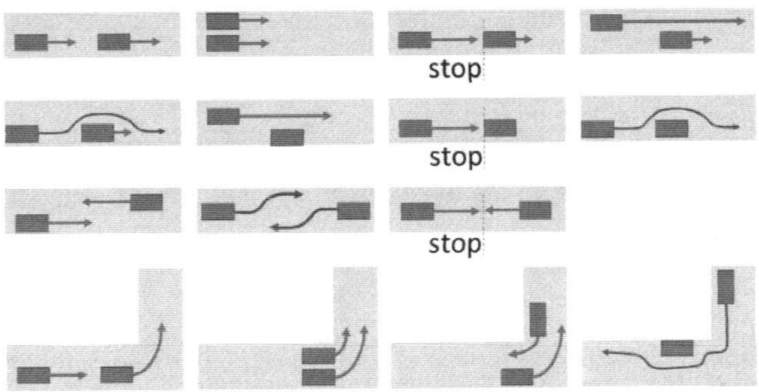

Figure 2a. Kinds of maneuvers (pair vehicles) were tested.

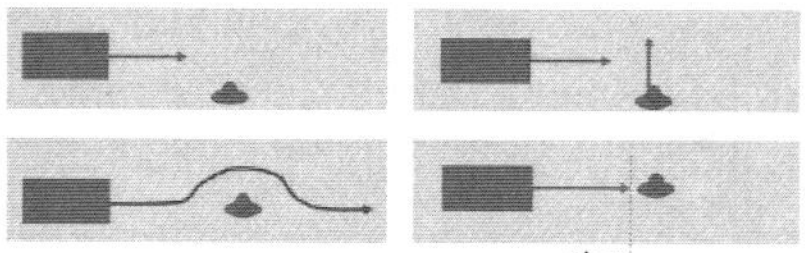

Figure 2b. Kinds of maneuvers (vehicle and pedestrian) were tested.

Experiments were conducted in July 2019 at private road in suburb of Moscow (Fig. 3)

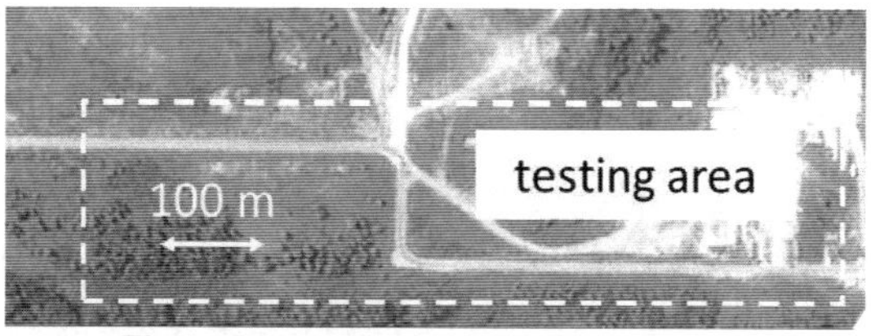

Figure 3. Testing area: road with two turns.

After series of experiments was counducted the written log-files were processed with open-source RTKlib software. Results are presented in the next section.

Models and algorithms

To estimatie the base vector the Least Square Method is used. The illustration of the mutural positioning is shown on fig. 4.

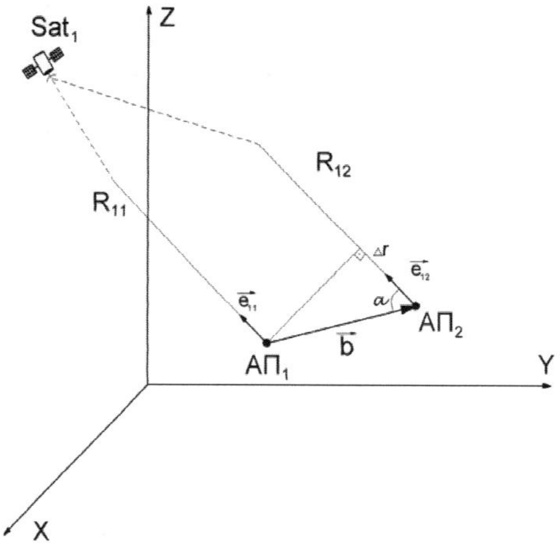

Figure 4. Scheme of mutual positioning.

Measurents in considered model are differences of pseudoranges:

$$\Delta\hat{r} = |\Delta\hat{r}_1, \dots, \Delta\hat{r}_N|^{\mathrm{T}}, \quad (1)$$

$$\Delta r_{\mathrm{j}} = R_{j2} - R_{j1}. \quad (2)$$

Relation between measurement and base vector:

$$\Delta r_j = (\vec{e}_{ji}, \vec{b}) = e_{ji_x} b_x + e_{ji_y} b_y + e_{ji_z} b_z. \quad (3)$$

System of equations:

$$\Delta\hat{r} = f(\boldsymbol{b}) = \begin{cases} \Delta r_1 = e_{11_x} b_x + e_{11_y} b_y + e_{11_z} b_z \\ \vdots \\ \Delta r_N = e_{N1_x} b_x + e_{N1_y} b_y + e_{N1_z} b_z \end{cases}, \tag{4}$$

where N is amount of visible satellites.

The design matrix can be written as:

$$\Delta \frac{\partial f(\boldsymbol{b})}{\partial \boldsymbol{b}_0} = H(\boldsymbol{b}) = \begin{vmatrix} e_{11_x} & e_{11_y} & e_{11_z} \\ \vdots & \ddots & \vdots \\ e_{N1_x} & e_{N1_y} & e_{N1_z} \end{vmatrix}. \tag{5}$$

For esimation the iteration algorythm is used:

$$\boldsymbol{b} = \boldsymbol{b} + (\mathbf{H}^{\mathrm{T}}\mathbf{W}^{-1}\mathbf{H})^{-1}\mathbf{H}^{\mathrm{T}}\mathbf{W}^{-1}\boldsymbol{\rho} \tag{6}$$

where **W** is covariance matrix of observations.

The effect of a pseudodelay smoothing filter by pseudophase measurements was also investigated. The filter algorithm is described below:

The measurement model:

$$\boldsymbol{y}_k = \boldsymbol{C}\boldsymbol{x}_k + \boldsymbol{n}_k \tag{7}$$

where $k = 0,1,2,\ldots$ is discrete time; $\boldsymbol{n}$ is vector of discrete white gaussian observation noise with independent components.

The state vector includes pseudodelay, rate of change of pseudodelay and pseudophase:

$$\boldsymbol{x}_k = [\tau\ \dot{\tau}\ \varphi]'. \tag{8}$$

Matrix **C**:

$$\boldsymbol{C} = \begin{pmatrix} 1 & 0 & 0 \\ 0 & 0 & 1 \end{pmatrix}. \tag{9}$$

Dynamic model:

$$\boldsymbol{x}_k = \boldsymbol{F}\boldsymbol{x}_{k-1} + \boldsymbol{G}\gamma_{k-1}. \tag{10}$$

where γ is discrete white gaussian dynamic noise.

Matrices **F** and **G**:

$$\boldsymbol{F} = \begin{pmatrix} 1 & T & 0 \\ 0 & 1 & 0 \\ 0 & T/\lambda & 1 \end{pmatrix} \boldsymbol{G} = \begin{pmatrix} 0 \\ T \\ 0 \end{pmatrix}. \tag{11}$$

where T is discrete time's period; λ is GNSS's wavelength.

Experiments

Figures 5-7 show sample results from 15 ongoing experiments. Table 1 shows the values of the errors calculated relative to the reference trajectory. Comparisons are made between estimates obtained using smartphones in stand alone mode (stand alone), estimates obtained in relative mode without smoothing pseudo-ranges (without smoothing), and estimates obtained using a filter that smooths pseudo-ranges (with smoothing). Consider three paths:

1. Riding cars one after another and turning (fig. 5)
2. Overtaking of one car by another (fig. 6)
3. Passage of one car past standing (fig. 7)

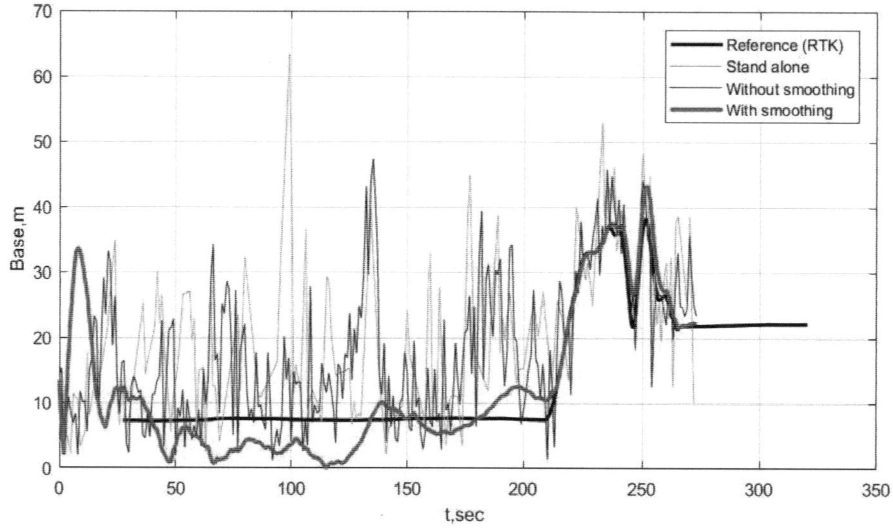

Figure 5. Baseline estimations for path 1.

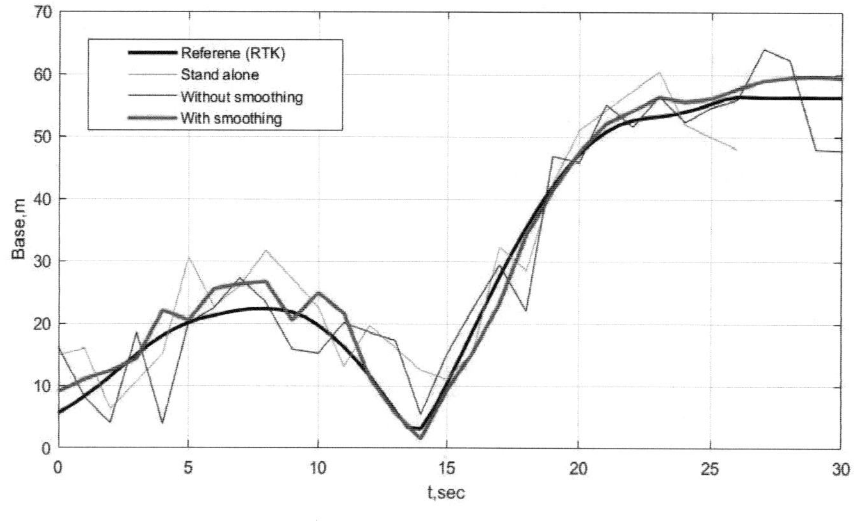

Figure 6. Baseline estimations for path 1.

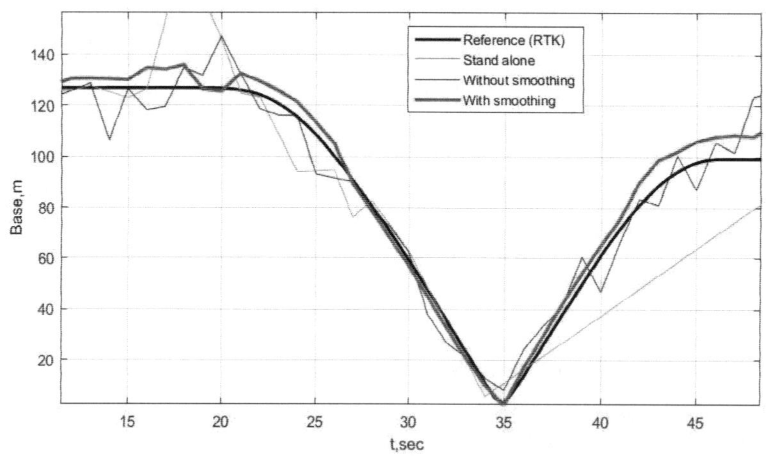

Figure 7. Baseline estimations for path 3.

Table 1. Experiment results

Path number	Standart deviation, m		
	Stand alone	Without smoothing	With smoothing
1	11.2	8.83	3.39
2	6.15	5.44	2.48
3	17.1	10.5	4.3

The stand alone solution for two smartphones provided the highest error of baseline estimation, and the smallest error was provided in the mutual navigation mode with smoothing pseudorange using measurements of the pseudophase, it is on average 3 times less than that achieved in stand alone mode and 2 times less than that achieved in mutual mode without smoothing.

Discussion

Experiments show that access to raw GNSS measurements in driver's smartphone and its joint processing provides opportunity to track mutual trajectories of vehicles with much less uncertainty than general usage of smartphone. In some cases its quality is enough for direct collision warning. But there are a lot of cases whith high uncertainty of estimation of mutual trajectories which is not enough for reliable collision warning. However there are potential fields of improving reliability of trajectory estimation and prediction: trajectory filtering, integration with smartphone's in-built inertial sensors, wireless integration with on-board sensors (speedometer, steering angle), matching with 3D-map of city and duty cycle turning off.

Trajectory filtering is based on knowledge of regularity of vehicle's trajectory and provide effective smoothing. Integration with inertial sensors allows to narrow effective bandwidth of trajectory smoothing filter to futher reduce noise error of mutual positioning simultaneously reducing dynamic error. Integration with on-board sensors allows fast and reliable adaptation of trajectory filter getting moments of static modes and steer's turns. Matching with 3D-map allows to implement software mask to cut some damaged signals. Turning off the duty cycle provide much better quality of tracking pseudoranges and cariers' psedophases; using smartphone in vehicle could be easily conducted with external power supply, so duty cycle is not necessary for in-car driver assist use case.

Implementing the noted features we're going to provide sub-meter uncertainty of mutual inter-vehicle trajectory tracking and collision warning with high reliability enough for practical implementation.

Conclusion

Smartphone is powerfull tool for wide variety of activities. Recent years modern Android smartphones obtained possibility of pricision GNSS positioning. Given wide spreading among people and new opportunities smartphones could improve road safety by warining drivers about probable danger of collision.

Experiments showed that in ideal conditions the best smartphones provide sub-decimeter uncertatinty. But real-life conditions limit uncertainty to level of about 1..3 meter. Experiments with vehicles and pedestrian showed that direct processing of raw GNSS measurements from smartphones don't provide reliable enough warning.

However there are additional opportunities for improving of collision warining reliability using access to raw GNSS measurements, their sharing and joint processing.

Our future work in focused on trajectory filtering, integration with smartphone's in-built inertial sensors, wireless integration with on-board sensors (speedometer, steering angle), matching with 3D-map of city and duty cycle turning off.

First results using smartpones are promissing and we're going to investigate efficiency of more complex processing for collision warining.

References

[1] F. de Ponte Müller, "Survey on ranging sensors and cooperative techniques for relative positioning of vehicles," Sensors, vol. 17, no. 2, p. 271, 2017

[2] Simon Banville, Frank van Diggelen - Precise GNSS for Everyone: Precise Positioning Using Raw GPS Measurements from Android Smartphones. GPS WORLD 2016, 27, 43-48

[3] Humphreys, Todd E., Murrian, Matthew, van Diggelen, Frank, Podshivalov, Sergei, Pesyna, Kenneth M., Jr.,, "On the Feasibility of cm-Accurate Positioning via a Smartphone's Antenna and GNSS Chip," Proceedings of IEEE/ION PLANS 2016, Savannah, GA, April 2016, pp. 232-242

[4] Scott Stephenson, Xiaolin Meng, Terry Moore, Anthony Baxendale, and Tim Edwards. Network RTK for Intelligent Vehicles - GPS World, February 2013

[5] Simon Banville, Frank van Diggelen - Precise GNSS for Everyone: Precise Positioning Using Raw GPS Measurements from Android Smartphones. GPS WORLD 2016, 27, 43-48

[6] Pirazzi, Gabriele, Mazzoni, Augusto, Biagi, Ludovico, Crespi, Mattia, "Preliminary Performance Analysis with a GPS+Galileo Enabled Chipset Embedded in a Smartphone," Proceedings of the 30th International Technical Meeting of the Satellite Division of The Institute of Navigation (ION GNSS+ 2017), Portland, Oregon, September 2017, pp. 101-115.

[7] Lu, Yangwei, Ji, Shengyue, Chen, Wu, Wang, Zhenjie, "Assessing the Performance of Raw Measurement from Different Types of Smartphones," Proceedings of the 31st International Technical Meeting of the Satellite Division of The Institute of Navigation (ION GNSS+ 2018), Miami, Florida, September 2018, pp. 304-322.

[8] Warnant, René, De Vyvere, Laura Van, Warnant, Quentin, "Positioning with Single and Dual Frequency Smartphones Running Android 7 or Later," Proceedings of the 31st International Technical Meeting of the Satellite Division of The Institute of Navigation (ION GNSS+ 2018), Miami, Florida, September 2018, pp. 284-303.

[9] Laurichesse, Denis, Rouch, Cedric, Marmet, Francois-Xavier, Pascaud, Matthieu, "Smartphone Applications for Precise Point Positioning," Proceedings of the 30th International Technical Meeting of the Satellite Division of The Institute of Navigation (ION GNSS+ 2017), Portland, Oregon, September 2017, pp. 171-187.

[10] Sharma, Himanshu, Schütz, Andreas, Pany, Thomas, "Emulating the Smartphone GNSS Receiver to Understand and Analyze the Anomalies in RTK Positioning using GNSS Raw

[11] Measurements," Proceedings of the 2019 International Technical Meeting of The Institute of Navigation, Reston, Virginia, January 2019, pp. 577-582.

[12] Liu, Wei, Liu, Bingcheng, Chen, Xiao, "Multi-sensor Fusion Algorithm Based on GPS/MEMS-IMU Tightly Coupled for Smartphone Navigation Application," Proceedings of the 30th International Technical Meeting of the Satellite Division of The Institute of Navigation (ION GNSS+ 2017), Portland, Oregon, September 2017, pp. 2973-2980.

[13] Wu, Zida, Liu, Peilin, Liu, Qiang, Wang, Yuze, "MEMS-based IMU Assisted Real Time Difference Using Raw Measurements from Smartphone," Proceedings of the 31st International Technical Meeting of the Satellite Division of The Institute of Navigation (ION GNSS+ 2018), Miami, Florida, September 2018, pp. 445-454.

[14] Qiong Wu, Mengfei Sun, Changjie Zhou and Peng Zhang. Precise Point Positioning Using Dual-Frequency GNSS Observations on Smartphone - Sensors, May 2019

[15] Kurum, Mehmet, Gurbuz, Ali Cafer, Nelson, Claudia, Orsini, Lauren, Scheider, Mia, "On the Feasibility of Smartphone-based Interferometric GNSS Reflectometry," Proceedings of the ION 2019 Pacific PNT Meeting, Honolulu, Hawaii, April 2019, pp. 635-640.

[16] https://developer.android.com/guide/topics/sensors/gnss

Application of systems of telematics on commercial vehicles

H M Saleh [1], R V Nuzhdin [2], S V Kurochkin[3]

[1] Associate Professor of National Research University Higher School of Economics (NRU HSE), associate professor "Information systems and software engineering", Vladimir state university named after Alexander and Nikolay Stoletovs, 87 Gorky Street, Vladimir, Russia.
[2] Associate professor "Motor transport", Vladimir state university named after Alexander and Nikolay Stoletovs, 87 Gorky Street, Vladimir, Russia.
[3] Undergraduate of Institute of information technologies and radiotronics, Vladimir state university named after Alexander and Nikolay Stoletovs, 87 Gorky Street, Vladimir, Russia.

E-mail: bremsemaster@yandex.ru

Abstract
The analysis of systems of telematics of commercial transport is provided in article, the factors influencing its application ways of development of the systems "Fleet management" in Russia are offered and also the pilot model of telematic system, more available from the economic point of view, with function of forecasting of technical condition of the rolling stock on the basis of expert system is presented.

Keywords: Telematic system, commercial transport, expert system, system of onboard preliminary treatment of the vehicle, architecture of intellectual system.

The cargo transportation which is carried out by motor commercial transport constantly develops owing to what volumes of this type of transportations increase what statistics of the Ministry of Transport of the Russian Federation of [2] confirms. Owing to the development this field of activity has the expanded regulatory base, wide coverage zone and huge number of units of the rolling stock both private cargo carriers, and huge transport companies.

At implementation of cargo transportation the owner and/or the driver of single commercial car in view of requirements of the legislation existing in the Russian Federation faces need of implementation of the list of control operations:

- control of mode of behavior of drivers (onboard tachograph);
- control of observance of route (Platon system);
- control of weight condition of the vehicle (points of weight control);
- control of safety of freight (onboard preliminary treatment, in some configurations of vehicles);
- control of technical condition (onboard preliminary treatment, periodic technical inspection);
- documentary maintenance.

All above-mentioned operations can be realized or separately, using the specialized equipment, or with application of the complex "Fleet management" system including as telematics (remote control of the car in flight), and documentary maintenance.

The Fleet management systems are in our country at stage of development and have not received full application yet. At the moment in commercial motor transport the systems of telematics allowing depending on complete set are widely used, far off to obtain information on vehicle geolocation, fuel level in tank, technical condition by results of onboard preliminary treatment, etc. For descriptive reasons we will consider the analysis of the above-stated systems of various type (Tab. 1).

In fact, application of systems of telematics allows to resolve issue of improvement of control cargo system, but in our country it is necessary for further development of the sphere of "Fleet management":

- import substitution of systems of telematics of foreign production;
- improvement of consumer properties of system of remote preliminary treatment of commercial cars;
- expansion of universality, informational content and applicability of telematics.

The system which structure is presented in Fig. 2 can be one of versions of the solution of the sounded problems. The structure of system means following processes.

The Fleet management systems have similar structure (Fig.1).

1. Reading of parameters of technical condition of the vehicle braking system has to happen by means of connection to the regular socket (for example, like "OBD II") the reading-out adapter equipped with the Bluetooth radio-module, or "WiFi".

Table 1. Analysis of systems of telematics of comercial transport

Qualities and functions of system	**Name of telematic system**		
	«Transics» (WABCO) [5]	**«J-TDO» (Cojali Groop)** [4]	**«Truck-control»** [3]
Modularity	+	-	+
Geolocation	+	-	+
Control of technical condition of the tractor	-	-	-
Control of technical condition of the trailer	+	+	-
Control of weight state	+	+	+
Control of mode of behavior of the driver	+	-	+
Use of the server of data storage	+	-	+
Documentary maintenance	+	-	-
Cost of components	Does not function in Russia	≈30000 RUB	from 20000 RUB

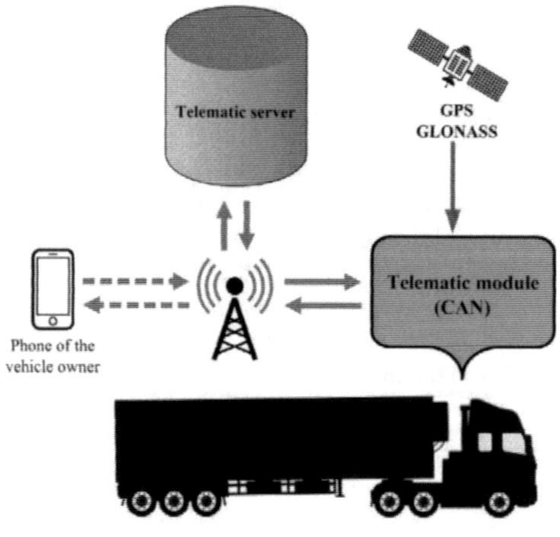

Figure 1. The developed architecture of system of Fleet management

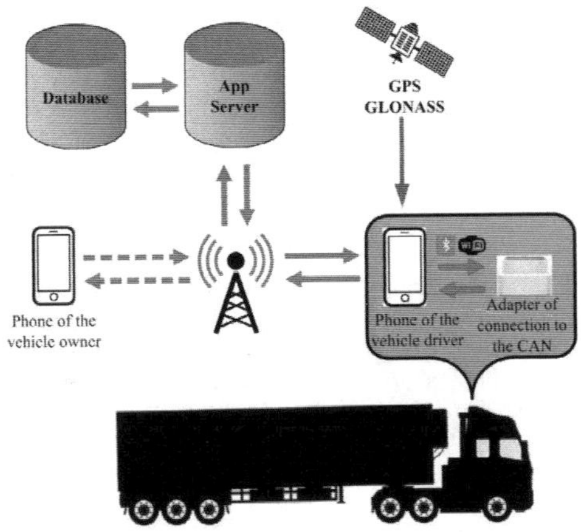

Figure 2. The offered system of telematics

2. Data storage of work of system of telematics has to be carried out on the special server.
3. Collecting, drive and display of data on technical condition of the braking system of the commercial car and its geolocation (above) and the installed developed application has to be realized by means of the smartphone (tablet) with the Android 4.0 operating system. Data transmission to phone from the reading-out adapter it has to be carried out under the Bluetooth protocol, or "WiFi". Data transmission on the server for storage of information and also on additional devices for remote control of technical condition, it has to be realized through Internet network. Phone has to obtain data on geolocation of the commercial car from the existing GPS and GLONASS networks by means of the built-in sensors.
4. Steering, the analysis and display of data on technical condition of the braking system of the vehicle and its geolocation is required to be carried out the developed software.
5. The developed system has to be autonomous, during its work the dispatching center will not be required. Application of this system has to allow to keep track of technical condition of the braking system of all road train.
6. Documentary maintenance of operations of transportation of freight has to be provided by integration into the software of system of the special module of document flow.

Main advantages of similar technical solution:

• complete solution for control of the vehicle in flight;
• exception of "human factor" in control operations by introduction independently trained expert system;
• possibility of improvement of the software and expansion of its functions;
• low cost of accessories of system;
• application of similar system does not demand modification of design of the car and installation of specialized telematic modules.

Besides, in this system use of expert system of onboard preliminary treatment with possibility of forecasting of technical condition of the car is offered.

The architecture of system of preliminary treatment of [1] failures of the vehicle represents two subsystems: subsystem of reception and information processing and subsystem of interpretation of the obtained information on condition of subject to diagnosing. The first carries out reception of data on condition of object and their further processing (distribution of data, assessment of variables and their display). The second subsystem makes recognition by means of artificial neural network of failures of systems of the vehicle, and the indistinct controller makes recommendations about realization of further actions (forms conclusions and the option proposes solutions).

Now IS are used at the solution of problems of various types: decision-making in the conditions of uncertainty (incompleteness), interpretation of symbols and signals, prediction, preliminary treatment, designing, planning, steering, control, etc.

The main design stages of model of neural network for assessment of technical condition of vehicle:

1) definition of the logical conditions allocating the analyzed selection in the registered transition processes;
2) task of symbolical indistinct variables for vectors of entrance and exit of the modelled object;
3) formation of the base of "indistinct" rules describing all possible operational modes of behavior of systems of the car;
4) creation on the basis of the registered selections of real transition processes of correct work of the training, testing and test data for training;
5) training of adaptive neuro and indistinct network.

Such representation allows to create the system of preliminary treatment on the basis of neural network, having defined all design stages. That is to create base of logical conditions and rules for system, to design it and to train.

Any system of preliminary treatment relies on probe of object of preliminary treatment, receiving and transformation of the signals arriving from object (interface), stay and malfunction repair and execution of the assumptions of further work of object.

On entrance of object of preliminary treatment the signal by means of which at the exit reliable information about work of all system in general is issued is given. The received signal via the interface will be transformed and arrives on entrance of neural network where according to operation algorithm of neural network it is investigated, and conclusions about condition of system are displayed on the screen. Thus, the problem of preliminary treatment of the electromechanical device on the basis of neuronets comes down to probe of changes of any parameter of the device, entering in the knowledge base of information on serviceability and failure of the device in the studied parameter, comparison of the put and obtained information and formation of offers on further operation of the device.

As it has been told above, one of properties of neural network is the ability to training, that is the expert-teacher, forming neural network, puts the knowledge, conclusions of serviceability or failure of this or that system in its knowledge base, and the neural network itself submits the conclusions about condition of system. It preliminary treatment considerably becomes simpler.

Neural networks also have the high computing power and high fail-safety.

At probe of influence of properties of neural network on quality of preliminary treatment on entrance of neural network the vector (matrix from two lines) with serviceable and faulty signals with vehicle moves. Reference of faulty points (aberration) to the second class (failure is result of training at change of parameters). Process of training is carried out and the network is tested for viewing of results of training. With increase in number of neurons the quality of recognition increases.

Let's consider in more detail the intellectual system of diagnosing of the braking system of the car. Its main objective is determination of technical condition of the braking system of the diagnosed car and issue of recommendations at misfit of operation parameters of this system to standard values. In this case the expert system "tree of decisions" is applied. Subtasks of system are:

1) Machine learning for improvement of operation algorithm.

2) Possibility of forecasting by results of collected data on work of system.

3) Determination of style of driving.

The solution of task is based on application of expert "tree" system. The settlement scheme of process of braking of the car is submitted in Fig. 3(a).

The accepted symbols:

- Θ_{Si} – the deflection of elastic element of mechanical suspender;
- p_{S1}, - pressure in air suspension of front axle;
- p_{S2} - pressure in air suspension of rear axle;
- S_{Bi} – the current value of braking distance;
- S_{Pi} – the current value of pressing brake pedal;
- G_1 - the weight of front axle;
- G_2 - the weight of rear axle;
- G_V – vehicle weight;
- V_{Wi} – speed has begun braking of wheel
- s_i – wheel slip;
- ε_{Vi} – the complex parameter of the braking system;
- j_{Bi} – the current value of the established slowdown when braking;
- j_E – reference value of the established slowdown when braking.

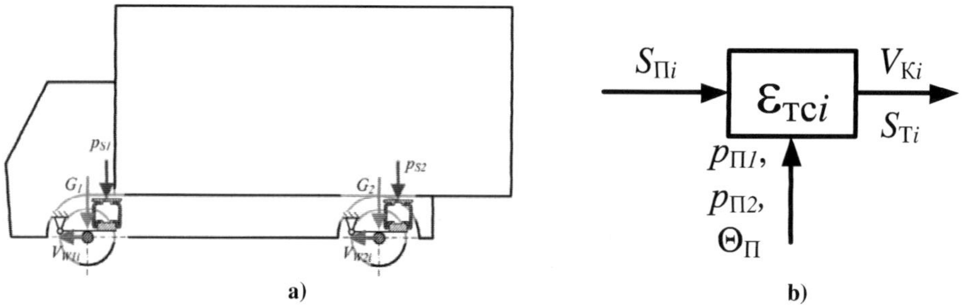

Figure 3. Model of braking of the vehicle:
a – the scheme of forces operating on car wheels; b – the settlement scheme of definition of diagnostic indicators.

The diagnosed system is considered as "black box" (see Fig. 3(b)) and its characteristics are described by the complex parameter of the ε_{TCi}. braking system. At braking implementation the system defines the current ε_{Vi} value and compares to reference and accounting of borders of admission:

$$\varepsilon_{Vi} \geq [\varepsilon_V] - \Delta([\varepsilon_V]);$$
$$\varepsilon_{Vi} \leq [\varepsilon_V] + \Delta([\varepsilon_V]).$$

During the work of expert system there is automatic correction of values $[\varepsilon_V]$ and $\Delta([\varepsilon_V])$ depending on wheel slip of s_i depending on weather conditions. In this case size $\Delta[\varepsilon_V])$ is criterion for evaluation of style of driving, the size $[\varepsilon_V]$ characterizes the current technical condition of the braking system. Proceeding from the level $[\varepsilon_V]$ the forecast of operability of the braking system is formed. Besides, during the work of expert system control of failures of the braking system is made ("X" parameter) at which repetition the system informs on need of urgent repair.

In Fig. 4 the architecture of intellectual system of diagnosing of the braking system of the car is presented, the algorithm of realization of expert system is given in Fig. 5.

At the initial stage of work collecting the database $[\varepsilon_V]$ and $\Delta([\varepsilon_V])$ depending on s_i for implementation of machine learning of expert system will be carried out.

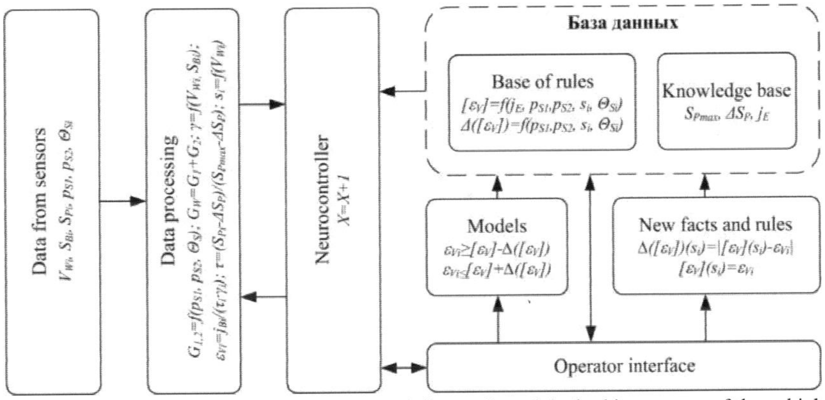

Figure 4. Architecture of intellectual system of diagnosing of the braking system of the vehicle

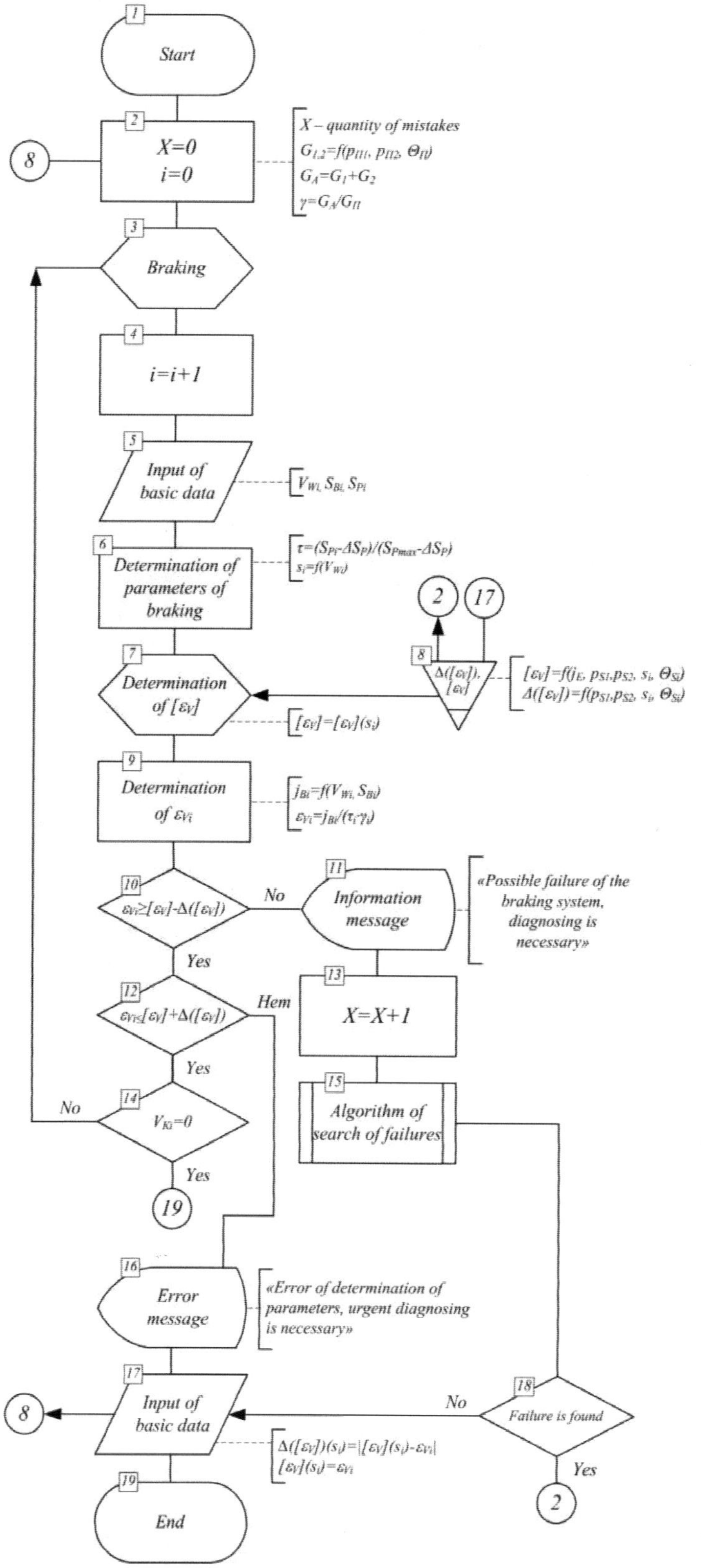

Figure 5. Algorithm of realization of expert system

The adjusting parameter - the level of loading of the vehicle (depending on suspender type - pressure in air suspension of front axle p_{S1} and pressure in air suspension of rear axle p_{S2}, or deflection of elastic element of mechanical suspender Θ_{Si}) and wheel slip of s_i.

The steering parameter - the current value of pressing S_{Pi}. brake pedal.

Output parameters - the speed of some wheel of V_{Wi} and the current value of braking distance of S_{Bi}.

Diagnostic parameters – value of complex parameter of the ε_{Vi} braking system.

Application of the presented system will allow:

- to increase availability of system of telematics of commercial transport;
- to exclude "human factor" at operation and forecasting of technical structure of commercial transport;
- to increase safety of transportations and traffic owing to timely prevention intellectual system about possible refusal;
- at introduction and development of the offered system there will be possibility of its integration with the systems of supervisory authorities that will lead to simplification of implementation of control operations by them.

References

[1] Veselov O. V. Metody iskusstvennogo intellekta v diagnostike : manual / Vladimir: Vladimir state university named after Alexander and Nikolay Stoletovs 2015

[2] TRANSPORT ROSSII. Informacionno-statisticheskij byulleten' 2018 URL: https://www.mintrans.ru/documents/7/9738

[3] Sputnikovyj GPS/GLONASS monitoring Truck-control 2019 URL: http://rus.truck-control.ru/

[4] Jaltest J-TDO Campo de Criptana, Ciudad Real, España 2019 URL:https://www.cojali.com/ru/%D0%9D%D0%BE%D0%B2%D0%BE%D1%81%D1%82%D0%B8-%D0%A1%D0%BE%D0%B1%D1%8B%D1%82%D0%B8%D1%8F/31/%D0%B1%D1%80%D0%BE%D1%88%D1%8E%D1%80%D0%B0-jaltest-j-tdo/

[5] Transics a WABCO compoany TRANSICS INTERNATIONAL BVBA Ieper Business Park - Zone K - Ter Waarde 91 - 8900 Ieper – Belgium 2019 URL: https://www.transics.com/

Multicriteria optimization in the study of active vehicle safety.

A.I. Markovnina, V.S. Makarov, V.V. Belyakov

Nizhny Novgorod State Technical University n.a. R.E. Alekseev, Minin st., 24, 603950, Nizhny Novgorod, Russia

E-mail: alinomalino@yandex.ru

Abstract. Ensuring road safety is a paramount task to be solved by specialists in the field of automobile transport. It affects all aspects of the Driver-Car-Road-Environment system and is aimed at increasing the degree of protection of road users from traffic accidents, at reducing the degree of road traffic injuries and reducing the risk of mortality. Similar functions for ensuring road safety are performed by the active safety systems of the car, including intelligent driver assistance systems and assistants. With all the variety of possibilities of equipping automobiles with complex systems, there is a need to assess the feasibility and profitability of installing a particular complex of systems. For this, it is proposed to apply multi-criteria assessment methods. At the same time, the possibility of multi-parameter optimization according to two or more criteria is considered, respectively, several approaches to solving the problem are described. As a result, a concept is developed that is close to universal, in assessing the active safety systems of a car and intelligent systems.

With all the variety of possibilities of equipping cars with complexes of systems, there is a need to assess the feasibility and profitability of installing a package. In this case, it is necessary to turn to multicriteria assessment methods.

The task is to determine the best of the options for complexes of active safety systems installed on a car. At the beginning, you must select one or more criteria by which the comparison and selection of options will take place. The variety of criteria is not limited to the list given in this article. For example, this can be the degree of prevention of road accidents as such, the degree of reduction in mortality, the degree of reduction in road traffic injuries, the degree of coverage of the variety of traffic situations leading to road accidents, the cost of equipping a car with a system, the cost of repair and maintenance of systems. At this stage, a list is formed: the system and performance indicators.

At this stage of the multicriteria assessment, a general indicator of the complex is formed, based on the assumption that there is a certain probability that at the onset of a pre-emergency situation at least one of the systems will work and prevent an accident / have time to reduce the speed of the car / will allow a safe maneuver.

After defining the criteria for comparison, it is necessary to choose a methodology for assessing, as well as a method for determining and assigning weighting coefficients for each criterion, if necessary. When considering two comparison criteria, the Edgeworth-Pareto set is constructed for clarity. The problem of extracting the Edgeworth-Pareto set is usually considered as a preliminary one. It is followed by the most significant stage of decision making [2]. An example of the distribution of alternatives when compared according to two criteria by the Edgeworth-Pareto method is presented in Fig. 1.

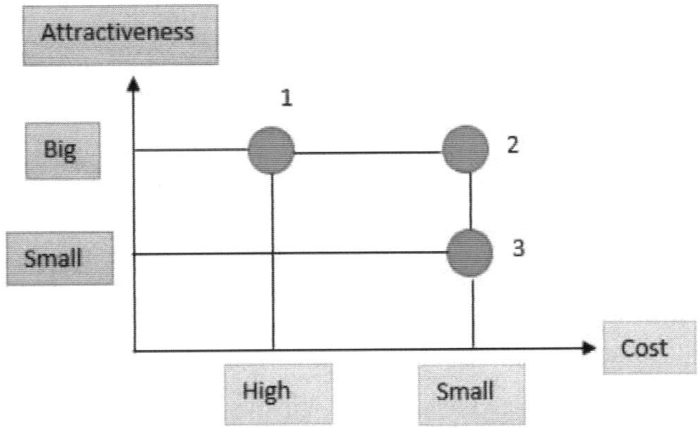

Figure 1. An example of the distribution of alternatives by the Edgeworth-Pareto method.

For a larger number of criteria, it is proposed to use the construction of scales of criteria [3]. In Fig. 2. presents a comparison of three models, evaluated by four indicators.

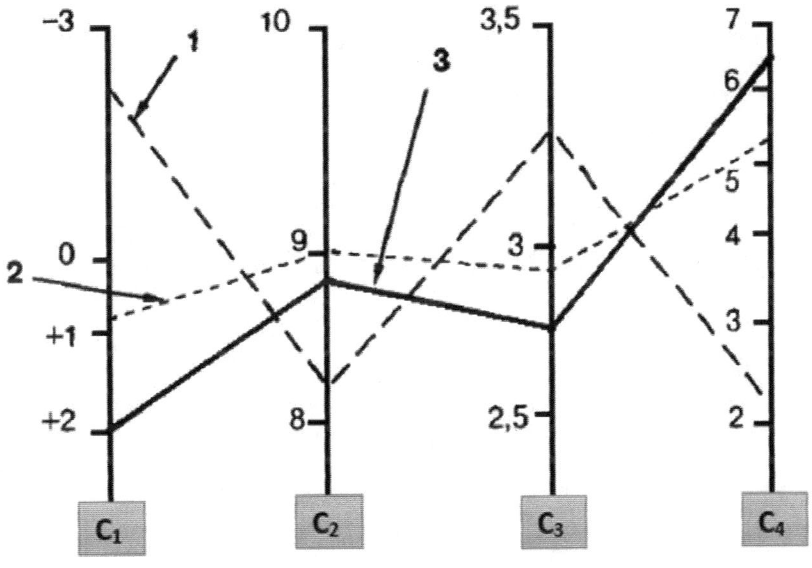

Figure 2. An example of constructing models according to criteria scales.

The cost of active safety systems is taken from authorized car dealers. If there are several prices for the same system, you can use different calculation options: maximum indicator, minimum or average. It is more expedient to take the maximum indicator when considering the multicriteria assessment problem in the long term, the average indicator in the short term, and the minimum when calculating at a given time.

When using some methods, the presence of weight coefficients is necessary, which can be determined in the following way.

The simplest method for determining the weighting coefficients is the method of direct assessment by one expert who selects the most important criterion, in his opinion, and gives him an assessment Ci=1, and for other criteria $C_i<1$ depending on the importance of each criterion. The disadvantage of this assessment is too subjective, because it depends on the opinion of one person. In this regard, often use a group of experts whose opinions are averaged, so the assessment is more subjective. Suppose there are m experts and n criteria, the rating set by the i-th expert to the j-th criterion is denoted by c_{ij}, where i = 1, 2, ..., m, j = 1, 2, ..., n. It is necessary to find the values of a_i based on expert estimates c_{ij}. If experts use the same scales in their assessments, then the values of a_i are calculated by simple averaging - dividing the sum of the estimates by the number of experts [4]. If different experts use different scales, then these estimates need to be normalized first. This is necessary in order to bring the rating scales used by different experts to one scale, in which all ratings are in the range from zero to one. Then the normalized expert estimates a_{ij} are calculated by the formula:

$a_{ij} = \frac{c_{ij}}{\sum_{i=1}^{k} c_{ij}}$, where

$\sum_{i=1}^{n} a_{ij} = 1$ - scaling condition.

The final values of the weight coefficients a_i are obtained by averaging the normalized estimates of all m experts for each particular criterion.

$$ai = \frac{\sum_{j=1}^{m} aij}{m}$$

The determination of weights is not limited to the proposed method.

The following describes the approximate calculation procedure using the example of the criterion "Probability of accident prevention".

Let us designate the type of accident "collision" - event 1, "collision" - 2, "rollover" - 3. The complex of AX_n systems will prevent event 1 with probability AY_n, the complex BX_n - event 2 with probability BY_n, the complex CX_n - event 3 with probability CY_n, where n is the serial number of the considered system or complex of systems.

If one system is considered, then there are no problems with it and the likelihood of accident prevention is simple and understandable. If the complex is considered, it is necessary to calculate the probability that at least one system will work in the pre-emergency or during an emergency.

Suppose P (A) is the degree of accident prevention by a complex of systems, P (A_n) is the reduction in the probability of accidents when using one system, P ($\bar{A}_n$) is the probability that the system n will not work properly.

$$P(\bar{A}n) = 1 - \mathrm{P(An)}$$

$$\mathrm{P(A)} = 1 - \mathrm{P(A1)} \times \mathrm{P(A2)} \times \ldots \times \mathrm{P(An)}$$

After calculating the probabilities, we choose the best options AX, BX and CX. To do this, we choose the comparison criteria and methods of multi-criteria evaluation. In the example, we consider the choice according to two criteria: the degree of accident prevention, calculated earlier, and the cost of installing complexes on the car.

Assign: the complex AX_n prevents accidents with probability P (A) = AY_n and costs S_{AXn}, the systems BX_n have P (A) = BY_n and cost S_{BXn}, respectively CX_n - P (A) = CY_n, S_{CXn}.

In this case, according to two criteria, one can construct the Edgeworth-Pareto set. Also, by assigning weight criteria to the method of expert evaluations, the following calculation can be done.

Suppose the weight of the cost criterion for installing W_S systems, the weight of the security criterion W_Y. Then, for each option, we calculate the utility function F by the formula:

$$F = S \times W_S + P(A) \times W_Y$$

Based on the results of calculations and graphing, the final complexes of systems AX, BX, CX of the corresponding cost are selected. The composition of the complexes determines the set of systems X, which is optimal and appropriate for installation on a car and to ensure road safety.

After determining the AX, BX, CX complexes, based on the statistics of accidents with victims during the January 3rd to February 26th period, one can calculate how many accidents could theoretically have been avoided if all the vehicles involved in the accident were equipped with active safety systems. This requires the total number of accidents by type.

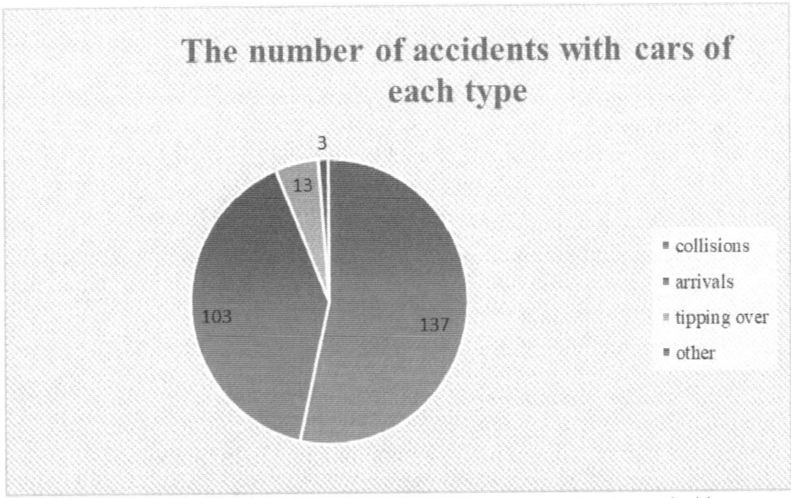

Figure 3. The distribution of the number of accidents for the period by type.

Table 1. The number of accidents of each type for the period.

collisions	139
arrivals	107
tipping over	13
other	11

In the presence of AX systems in event 1 (accident - collisions):

$N'_1=AY\times N_1$, where

N'1 – number of collisions prevented;

AY – probability of accident prevention by the AX system complex;

N1 – the number of accident type "collision".

With BX systems in event 2:

$N'_2=BY\times N_2$, where

N'2 – the number of accidents prevented such as a "collision";

BY - probability of accident prevention by a complex of BX systems;

N2 - number of traffic accidents of the "collision" type.

With CX systems in event 3:

$N'_3=CY\times N_3$ where

N'3 – number of collisions prevented;

CY - probability of accident prevention by a complex of CX systems;

N3 - the number of accident type "collision".

Further, the proportion of all types of accidental accidents prevented is considered K.

$K=(\sum N')/(\sum N)$, where

$\sum N'$ – the number of theoretically prevented accidents of all types;

$\sum N$ - the number of all accidents for the January 3rd to February 26th period.

Next, we examine the economic feasibility of installing the selected complex of X systems. For this, the cost of installation or the cost of installing and repairing intelligent systems is compared. It is worth mentioning the urgent problem of assessing the cost of human life and the cost of injuries.

The problem of determining the cost of human life is one of the most controversial in the field of automobile transport. The cost of human life, or the cost of the average life, is a conditional estimated economic value, since human life is not a market commodity.

There are objective and subjective approaches to assessing the value of human life.

Objective assessment methods are called methods that rely on the following performance indicators:

· Per capita lifetime income of a person, or total lifetime income;

· Average GDP per capita;

· Total lifelong income and average per capita GDP, adjusted for the likelihood of death at a particular age;

· Social costs of human reproduction;

· The balance of material goods and services consumed and accumulated by a person.

The above approaches take into account economic losses only after a person dies. But there are cases when a person survives, but loses the ability to work in general or only to certain types of work. Considering the area of road transport and the cost of accidents, it is more convenient to use the methodology by which insurance insurance payments are computed. In turn, insurance companies are subject to the Federal Law "On Compulsory Third Party Liability Insurance of Vehicle Owners" N 40-FL.

To do this, two indicators are compared: the cost of injuries of all victims of road accidents during the January 3rd to February 26th period and the cost of installing systems on all cars involved in these accidents. Here you can evaluate both the current situation and the situation in the future, given the life of the car, the cost of periodic maintenance of the systems and the number of accidents that can occur. Ideally, the cost of injuries should be higher than the cost of equipping a passenger car with systems. But in the accident reports on the website of the State Traffic Safety Inspectorate for the Nizhny Novgorod Region, accident details are only described for the dead and wounded, and road accidents with material damage are only numerically mentioned, which gives reason to conclude that if the conditions for the economic feasibility of installing the systems described above are not met, it is impossible to conclude that this action is completely inappropriate.

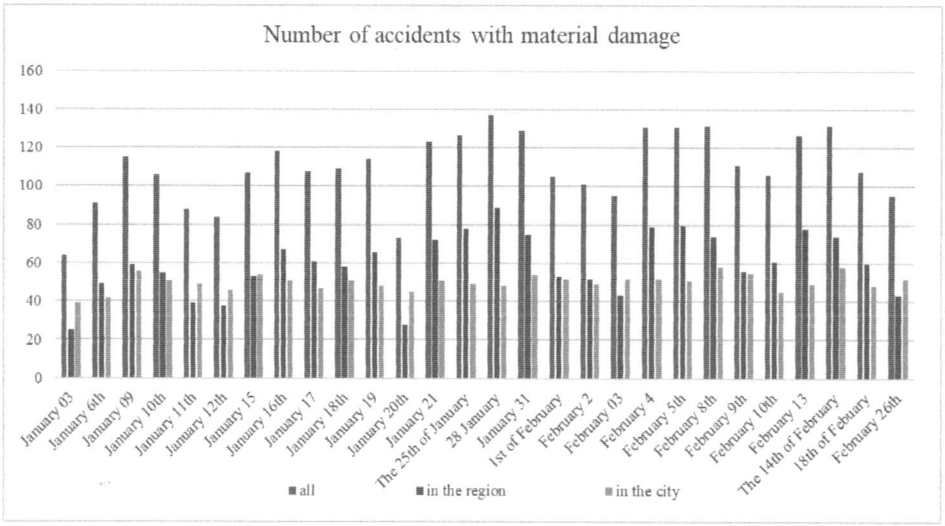

Figure 4. Number of accidents with material damage

The multicriteria optimization method gives clear conclusions, but does not guarantee one clear solution, the best in all respects at once.

IOP Conf. Series: Materials Science and Engineering **695** (2019) 012015 doi:10.1088/1757-899X/695/1/012015

References

[1] Automatic and intelligent vehicle systems. Cars and tractors, multi-purpose wheeled and tracked vehicles, ground transportation and technological complexes, mobile robots and planet rovers: a textbook / under the general. Ed. Professors V. Belyakov and L. Palkovich; Nizhny Novgorod State Technical University R.E. Alekseeva. - N. Novgorod, 2012 .-- 475 p.
[2] Larichev O. I. Theory and decision-making methods, as well as the Chronicle of events in the Magic Countries: Textbook. - M .: Logos, 2000.229 s: silt
[3] Lomakin V.V. et al. Vehicle safety. Textbook for high schools. - M .: MSTU "MAMI", 2011. - 299 p.
[4] Chernorutsky I. G. Methods of decision making. - SPb .: BHV Petersburg, 2005 .-- 416 s: silt.

Monitoring systems development using situational approach and fuzzy logic

Ermilov A.E.

Department of Computer systems and technologies, Nizhny Novgorod State Technical University named after R.E. Alekseev, Institute of radio-electronics and informational technologies, Nizhny Novgorod, Russian Federation

E-mail: aermilov09@mail.ru

Abstract
The article describes trends of the designing modern monitoring systems: scenario and contingent approaches. The authors propose to use the frame model to represent operation the systems. The frame model supports the using of fuzzy logic.
Keywords: frame machine, trends of using frame model, fuzzy logic.

Introduction

In given paper we consider top trends in subject domain of monitoring systems development.

The trend of intellectualization is traced in works [1-7]. Intellectualization in the narrow sense means active use of intelligent sensors which not only obtain information on object of monitoring, but also are capable to process it, are capable to the analysis of external/internal environment, to adapt to changes, to use difficult procedures of self-testing, calibration, etc.

Intellectualization in the broadest sense means active use of the AI for drawing up domain models, for processing of the arriving information and the analysis of the saved-up statistics: transition from data to knowledge – identification of the hidden dependences, the forecast of the situation, etc.

Intellectualization is not the purpose, but means of achievement of the goal – construction of flexible monitoring systems. Intellectualization allows not only to consider specifics of data domain of creation of monitoring systems, but also to create tools to display the situation in the form of an interactive image and the elements of emotional coloring of the situation that facilitate the work of the AWP operator.

Integration trend. From definition of monitoring systems it follows that functions of monitoring can be assigned to one of parts (subsystems) of any ACS, or can represent the independent system. Today it is obvious that the monitoring system can be the object of both "horizontal", and "vertical" integration.

In papers [8-10] it is noted that functions of monitoring are performed by subsystems in automated systems. The trend of development of monitoring systems which are oriented to the integration with other monitoring systems and managements is observed. Thus, when developing the monitoring system it is reasonable to be guided by already existing standards (GOST-34, IEEE 29148-2011) of interaction between the ACS elements.

Good basis of process of integration is the unified theoretical basis of association of different systems: for example, the use of the situational approach for development of monitoring and control systems. It seems promising to build a technology (based on the situational approach) of software and hardware monitoring systems using tools in the form of a system template, which can be tuned to a specific subject area reducing the cost of building the considered class of systems. A good basis for the integration process is a single theoretical basis for combining different systems: for example, the use of a situational approach to the construction of monitoring and control systems. One of the promising areas is the creation of technology for building monitoring systems for intelligent road transport systems, including highly automated software and hardware that can be customized to a specific subject area and purpose, reducing the cost of building such systems [11].

Theoretical analysis

Note that the monitoring systems solve a number of typical subtasks that are regularly repeated:

-check system performance, testing subsystems;
-identification of problems and emergency situations;
-parametric adjustment (calibration, etc.);
-monitoring of the work of subsystems (performance testing, etc.);
-system adjustment for changes in the external / internal environment of the monitoring system.

By "situation" we mean an object or process that is reproduced in the subject area of the monitoring system. The situation is described using a system of interrelated parameters. In the classical theory of frames, M. Minsky proposed using the network of frames [12] to describe a system of interrelated situations.

In [13], it was shown that the frame network, which describes the sequence of events in the scenarios of the interactive system, consists of three frame classes:
-frames-directive message;
-frames-dialog branching;
-frames-transformation of information on algorithms for the implementation of computational procedures.

Further development of this direction led to the description of the process of functioning of remote monitoring systems using a frame network, which included new classes. In paper [14], an extended frame unit was considered, the classification of which included a multimedia frame as the basis for building these monitoring environments.

As a result of the D.A. Belov's research, new prospects were identified for the development of frames in creating additional classes (frames, tools for building them) in the form of the "Multimedia Frame" class (MF). The subclasses are: MF - electronic map, MF - presentation, MF - diagram with the emotional coloring of the situation, MF - video, MF - animation.

The conceptual representation of the subject area of the monitoring systems operation allowed automating the creation of software through the creation of toolkits for each type of the frame.

The toolkits are combined in the form of a database of program code templates (or tool objects of the technology of visual software building). Libraries are equipped with tools that facilitate their maintenance during the life cycle, facilitate the "transformation" of the template into a program model, etc.

Thus, tools are created allowing to get a "guaranteed result" - a workable software code for machine processing of the situation in the subject area of the monitoring system.

Technique

As part of the trend described above, it is proposed to create specialized tools that facilitate the construction of monitoring systems.

To solve this problem, it is proposed to use specialized software modules that allow to process information using fuzzy logic procedures [15] (Fig. 1, 2). This allows you to complete well-known classes of frames, including conceptual descriptions of system work - frames with fuzzy logic procedures.

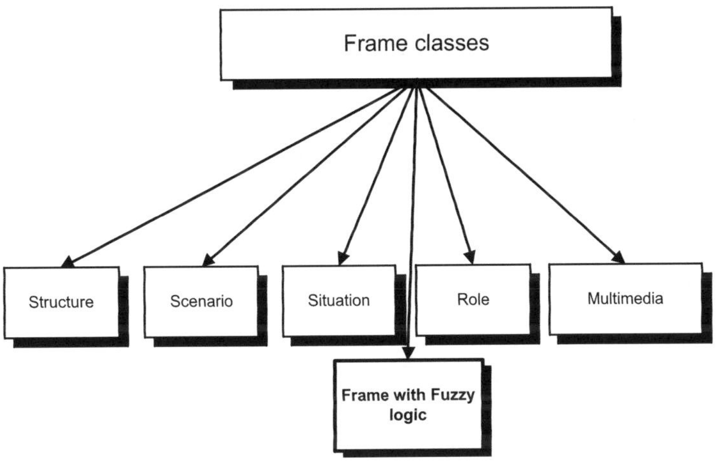

Figure 1. Extended frame classification

We introduce and consider such a concept as a frame with fuzzy logic (FFL). FFL will be understood as a frame, the slots of which are completed by fields using fuzzy logic (fuzzing values), as well as slot-results, the values of which are obtained by processing fuzzy inference when processing the initial values of other slots. The frame structure is shown in Figure 2.

idF					
idS	**T**	**D**	**V**	**L(V)**	**μ_L(V)**
Slot 1	Type 1	Daemon 1	value	Lex.Value	truth degree
Slot 2	Type 2	Daemon 2	value	Lex.Value	truth degree
• • •					
Slot N	Type N	Daemon N	value	Lex.Value	truth degree

Figure 2. Structure of frame with fuzzy logic

The type of slot values can be as numeric, as text values, and references to child frames (subframes). Frames with fuzzy logic can also be used as subframes. Storage of procedures with fuzzy logic in the frame is carried out in the form of lists of descriptors that are attached to the frame slots, and are executed when the values in the frame slots are updated.

To formalize this, we introduce the following frame designation $Fr = <idF, S, R>$, where:

- idF is the name or identifier of the frame;
- $S = <s_1, s_2 ... s_n>$ is the set of slots containing the facts defining the declarative semantics of the frame, where n is the number of these slots;
- $R = <s_1, s_2 ... s_m>$ is the set of slots that provide links to other frames (semantic, logical, etc.), where m is the number of these slots.

Frame slots can be formalized as follows:

- $s = <idS, T, D, V, L\ (V), \mu L\ (V)>$, where:
- idS is the identifier or name field (required);
- T is the slot type field (required);
- $D = <f_1, f_2, f_k>$ is a set of daemon procedures, including fuzzy logic, where f is an attached procedure descriptor, k is the number of frame procedures;
- V is a cell with a value, including subframe link;
- $L\ (V)$ is the lexical value field of the linguistic variable's term - the name of a fuzzy set according to the rule ;
- $\mu L\ (V)$ is the degree (measure) of belonging to the set of terms of a linguistic variable.

The basis for the implementation of fuzzy inference is the rule base, containing fuzzy "if-then" statements and membership functions for the corresponding terms of linguistic variables described earlier. The following conditions are required [16]:

Condition 1

There is at least one rule for each term of output linguistic variable (the sequential - the right part of the rule).

Condition 2

For any term of the input linguistic variable, there is at least one rule in which this term is used as a prerequisite (the antecedent - the left part of the rule).

The process of fuzzy inference is carried out in several stages (Figure 3):

1)development of fuzzy inference rules (pre-operative stage);
2)fuzzification (the stage of specifying the values of linguistic variables);
3)aggregation (the stage of determining the degree of belonging conditions for each of the rules of the fuzzy inference system);
4)activation (the stage of formation of membership functions of sequential production rules);
5)accumulation (the stage of finding the membership function of the output linguistic variable);
6)defuzzification (this is the process of transition from the membership function of the output linguistic variable to its precise numerical value).

In this paper, the Mamdani output algorithm is used. The defuzzification process implemented according to the center of gravity method [16].

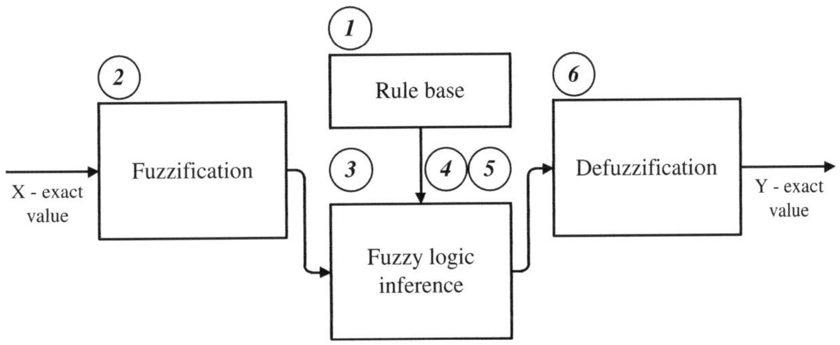

Figure 3. Fuzzy logic inference scheme

Let us consider the example of using the frame with fuzzy logic (Figure 4). Using fuzzy production rules, the Mamdani algorithm calculates the alarm level value for subframes in the frame model tree, which determines the order of output of information from these subframes, as well as the choice of executing scenarios connected with these subframes by condition (for example, sending alerts). The fuzzy logic apparatus allows to generate an alarm level (as the final value of the calculations for the frame) depending on several factors (a set of monitoring parameters). In this case, a human-oriented interface is used - verbal rules-statements to form the conditions for the presentation of information about the monitoring object.

Following example (“CPU” subframe) describes the dependence of the processor alarm level indicator for a computer server on temperature, in combination with the server load and the cooling fan speed. The rule base may contain the following statements:

1)If the processor temperature is “medium” and the server load is “small” and the fan speed is “high”, then the alarm level indicator is “high”;

2)If the processor temperature is “high”, then the alarm level indicator is “high”;

3)If the processor temperature is “low”, then the alarm level indicator is “low”;

4)If the processor temperature is “medium” and the server load is “small” and the fan speed is “low”, then the alarm level indicator is “low”.

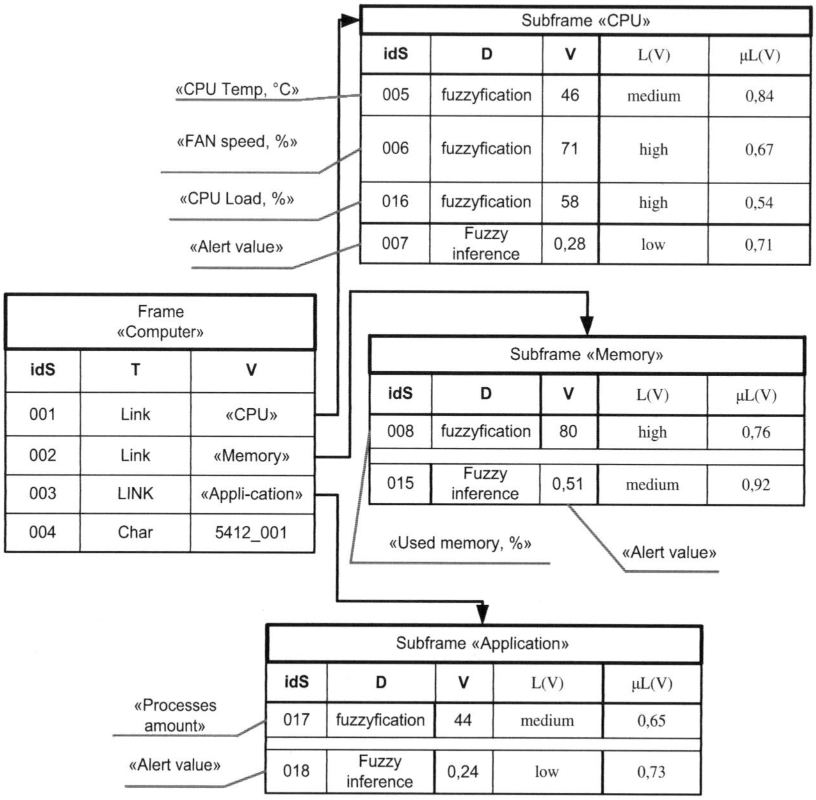

Figure 4. An example of building a network of frames with fuzzy logic

Let us consider the scheme of the system template (Figure 5).

Here are the subsystems of the developed monitoring system template. This scheme differs from the analog subsystem. The structure of the tool complex includes the following elements:

1. Sources of information. Various methods of collecting information about the monitoring object are supported. This may be a computer network from other monitoring system and database. In our case (computer monitoring), agent programs are used.
2. Database. This subsystem is a means of accumulating "raw" data that do not describe the situation in the subject area.
3. Bank of situations stores reference situations in the monitoring domain, which are presented in the form of a frame model.
4. Information analysis system - software for processing information about a monitoring object.
5. The user interface.

The main typical tasks of the subsystem are:

• monitoring results data;
• situational analysis;
• formation of a bank of reference situations;
• highlighting of duplicate elements, conflicts and their resolution in the bank of situations.

6. Interpreter processes raw data and forms a situational description from a raw data - a frame. It compares the situational description with the elements of the situational bank, assesses the situation at the monitoring facility and interacts with the user (monitoring operator), warning the person (if necessary) about negative trends in the development of the situation at the monitoring facility.
7. The statistical data processing system allows obtaining "traditional" characteristics of time series from raw data.
8. The logging system for recording the operation of the monitoring system facilitates the expansion and debugging of the instrumental complex.

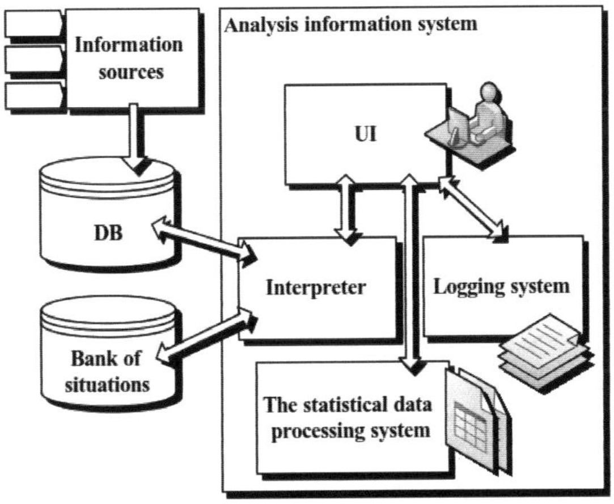

Figure 5. Generalized scheme of the monitoring system template

All subsystems of the monitoring system are functionally complete elements that are made in the form of skeletal shells that are tunable to a specific subject area (computing, transport, medicine, etc.). The skeletal monitoring system's shell is made in accordance with the integration trend, and supports the connection of new subsystems. For example, if a multi-user monitoring system is built, the connection of the administration system and the subsystem of the logging of user actions is supported. The apparatus of fuzzy logic was used in the frame model as procedures - daemons, which allowed in this configuration to implement a mechanism for informing the operator on priority importance (alarm level).

Conclusion

The article describes the tools for implementing a situational approach to the construction of monitoring systems. The process of describing the domain of the functioning of monitoring systems using a modified frame model is presented. The positioning of the proposed approach relative to current trends in the development of monitoring systems has been completed. The paper describes elements of typical architectural solutions for monitoring systems built on the basis of a situational approach, which creates prerequisites for the implementation of an instrumental complex for automating the construction of systems of this class. These approaches can be applied to the implementation of intelligent transport monitoring systems with a high level of automation to increase the level of reliability and ensure operational safety.

Acknowledgements

This research done with the financial support from Ministry of Education and Science of the Russian Federation (the unique project identifier is RFMEFI57717X0268). The experimental research conducted with the use of measurement equipment of the NNSTU Centre of collective using «Transport Systems».

References

[1] Kolosok I.N., Gurina L.A. Prediction of regime parameters in the monitoring and management of the power system. Electricity. 2014. № 1. P. 21-27.
[2] Shereshevsky L.A., Sidorov A.A. Methods to control the accuracy and reliability of data in distributed monitoring systems and automated control. Proceedings of the Samara Scientific Center of the Russian Academy of Sciences. 2002. Vol. 4. No. 1. P. 169-177.
[3] Vasyutin S.V. Making generalized decisions in the intrusion detection system, using several methods of monitoring data analysis. / Vasyutin S.V., Korneev V.V., Raich V.V., Sinitsa I.N. Information countering the threats of terrorism. 2005. No. 4. S. 54-65.
[4] Plyusnin I.I. Mobile system for diagnostic maintenance and monitoring of gas pipeline systems. / Plyusnin II, Bushmeleva K.I., Bushmelev P.E. / Fundamental research. 2006. № 1. S. 65.
[5] Chetyrbotsky A.I. Monitoring of states of multicomponent objects in geographic information systems. Pacific geology. 2006. V. 25. No. 5. P. 81-87.
[6] Zagvozdkin V.K. Information technologies for the preparation of control solutions in automated systems for the environmental monitoring of chemical industry enterprises. / Zagvozdkin V.K., Kolybanov K.Yu., Ravikovich V.I., Panova S.A., Chikovani M.A. / Protection of the environment in the oil and gas complex. 2008. № 10. S. 3-6.
[7] Meshcheryakova E.M. Monitoring of social tensions and implementation of social systems management functions. Bulletin of the Chuvash University. 2007. № 3. S. 340-346.

[8] Sazonov D.Yu., Vasilyeva V.V. The system of environmental monitoring as a subsystem of the intelligent transport system. In the collection: Organization of traffic and road safety in European cities materials of the International Youth Scientific and Practical Conference. Czech Technical University in Prague, FGBOU VPO "State University - UNPK". 2014. p. 46-49.
[9] Ryzhkov V.A. Development of a universal system for monitoring API functions of software systems based on the object model of components. In the collection: INFORMATICS: PROBLEMS, METHODOLOGY, TECHNOLOGIES materials of the XV international scientific-methodical conference. 2015. p. 367-370.
[10] Mitrofanov E.P. Informatization of monitoring changes in regional economic systems. Bulletin of the Chuvash University. 2007. No. 3. P. 442-445.
[11]Reinholtz. Team VictorTango's entry in the DARPA Urban Challenge. Journal of Field Robotics - JFR. 2009. №25. 125-162.
[12]Semenov, V.V. Principles of formation and fragments of knowledge base management theory. General mathematical software for computer-aided design / VV. Semenov.-M .: MAI, 1981. - 158 p.
[13]Minsky M. 1981, Framework for concurrent knowledge // D. Metzing ed. Frame conceptions and text understanding. - B .; N.Y .: Gruyter, 1980. 1-25.
[14]Belov, D.A. Problem-oriented automated system for monitoring train movement / D.A. Belov, P.V. Misevich, V.P. Hranilov // Automation in industry, 2009, №2. P.49-51.
[15]Ermilov, A.E. Application of frame model and fuzzy logic in the basis of building tools of automated monitoring systems / A.E. Ermilov, P.V. Misevich // Proceedings of the Nizhniy Novgorod State Technical University. R.Ye.Alekseeva, 2015, №1 (108), Nizhegor. state tech. un-t. - N.Novgorod, p.71-76
[16]A. Piegat, Fuzzy Modeling and Control, Springer, Heidelberg, Germany, 2001.

IOP Conf. Series: Materials Science and Engineering **695** (2019) 012017 doi:10.1088/1757-899X/695/1/012017

Designed gearbox reliability assessment techniques based on the real one

R.R. Yusupov, A.D. Yarzhemsky, A.V. Alexeyev, M.G. Korchazhkin, L.A. Berdnikov

Nizhny Novgorod State Technical University after R.Y. Alexeyev

E-mail: ro.iusupov@yandex.ru

Abstract. The paper shows the designed gearbox reliability assessment and analysis on the basis of the real one. The basic diagram has been analyzed as well as the units' eventual failures. Based on the designed structural diagram of the SU gearbox reliability level has been estimated. The measures to raise the reliability level have been proposed.

Key words: unmanned vehicle, reliability calculation, failure quota, mean time between failures, gear box.
Research: “Gazon Next” gear box reliability calculation.
Research objective: estimate the reliability of the designed gear box on the basis of the real one and update to the required reliability level.

Introduction

Reliability is the property of objects to keep in time the values of all the parameters describing the ability to function as required by the preset modes and operating conditions.

Reliability is dependable on the following factors:

- Road operation conditions;
- Transport operation conditions;
- Natural and climatic conditions.

Road conditions determine the car units' operation modes. They are determined by the road technical category, the type and quality of the roadway covering and the land topography (its altitude).

Transport operation conditions of the vehicle stock are determined by the traffic routing, cargo features, the real amount of load and others. All the above is determined by a number of basic rates and factors of vehicle operation conditions.

Natural and climatic conditions are characterized by the ambient temperature (a. t.), air humidity, dust level, precipitation intensity, wind load, solar radiation, seasonal fluctuations of operating conditions, environmental attacks, and altitude.

Car operation in urban and suburban areas causes time-varying modes of the ICE as well as intensive detrition of suspension, break gear and transmission rubbing parts, which are subject to heavy loads. [1]

At present non piloted vehicles are actively being developed. They require a high reliability level of their units and components. To achieve it computing and analyzing car units reliability level must be done, based on the obtained results, the reliability level must be raised to the goal, and new vehicle units must be designed.

“Gazon Next” gear box was chosen as the research object. “Reliability computing” and “updating to the goal” were made. Analysis was made based on the findings.

“Gazon Next” gear box reliability computing

Reliability computing plan

1. Gear box analysis.

2. Reliability structural diagram building.
3. Mean time to failure analysis.
4. System reliability computing.
5. System updating to reliability goal.

System basic diagram analysis

The general view of the basic diagram system is shown below (Fig. 1)

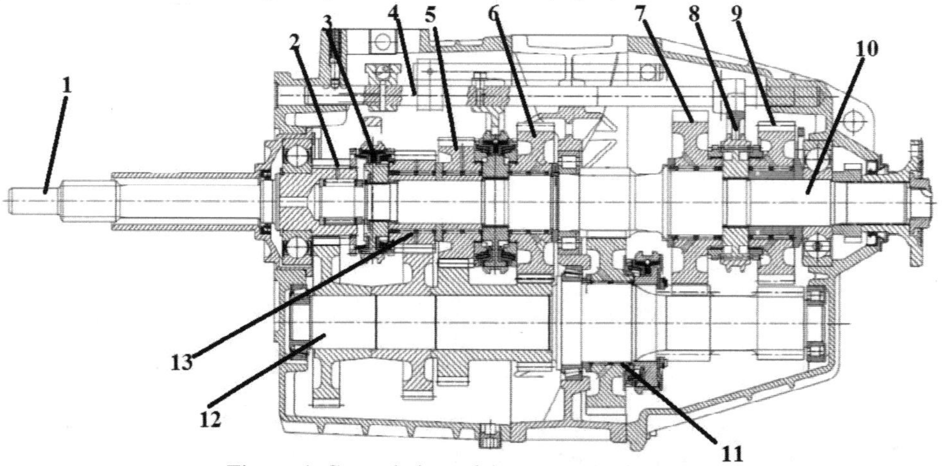

Figure 1. General view of the system basic diagram
1 – primary shaft; 2 – fifth gear; 3 – synchronizing sleeve; 4 – selector rod; 5 – third gear; 6 – second gear; 7 – first gear; 8 – gear shift fork; 9 – reverse gear; 10 – secondary shaft; 11 – sixth gear; 12 – intermediate shaft; 13 – fourth gear

Building structural diagram of the system reliability

Building is carried out with the following assumptions:

1) All the failures are independent;
2) In case a component fails, the system remains operative.

For building, let us subgroup the units based on their functionality:

1) gear shift knob
2) selector rod
3) gear shift fork
4) synchronizing sleeve
5) primary shaft
6) intermediate shaft
7) secondary shaft
8) gearwheel meshed

Let us make the system reliability model with the obtained units (Fig. 2)

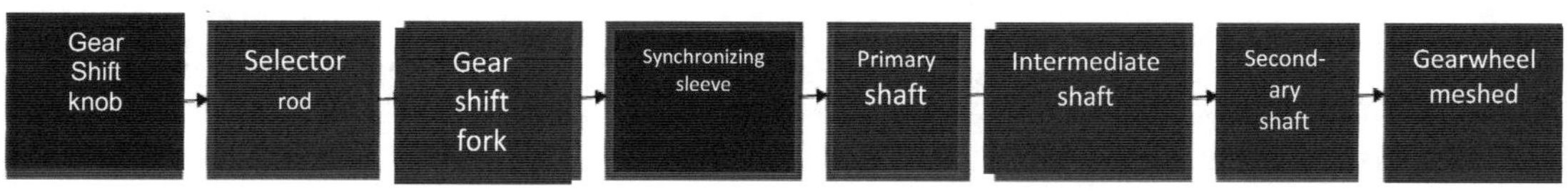

Figure 2. Air brake system reliability diagram

As the reliability of some units in the diagram is increased the reliability model has been enlarged (Fig. 3).

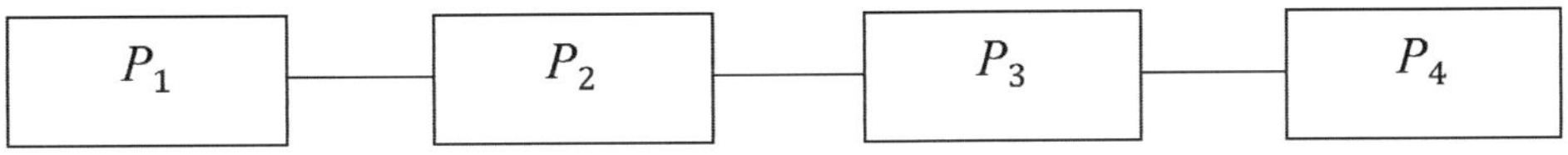

Figure 3. Enlarged reliability diagram

Mean time to failure analysis

Initial data array has been obtained from the OAO GAZ dealer stations information about the failure mean time (table 1)

Table 1 - Failure data

Units	Number of failures	Failure mean time km.	Reconditioning mean time norm/h.
GSN	1	94474	0.44
SR	3	51098	6.2
GSF	4	58544	6.2
SS	10	44996	5.99
Prim. shaft	3	51098	6.2
Interm. shaft	1	350	4.2
Secondary shaft	3	48522	5.92
Gearwheel meshed	4	39699	6.18

Fig. 4 shows the gearbox failure quota

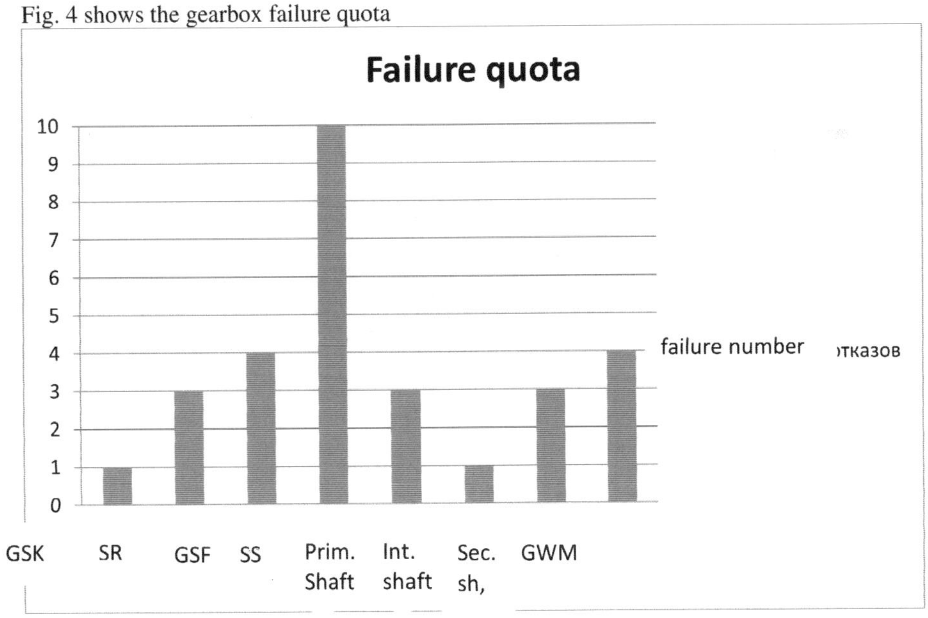

Figure 4. Units failure distribution

System reliability computing

Reliability computing algorithm is shown [2].

The reliability of this system is estimated:

$$P = P1 * P2 \bullet P3 * P4 * P5 * P6 * P7 * P8$$

where P_1 – Gear shift knob (0,96)
P_2– Selector rod (0,89)
P_3 – Gear shift fork (0,86)
P_4 – Synchronizing sleeve (0,64)
P_5 – Primary shaft (0,89)
P_6 – Intermediate shaft (0,96)
P_7 – Secondary shaft (0,89)
P_8 – Gearwheels meshed (0,86)

Probability of no-failure operation according to experimental data of the system failures is:

$$P = P1*P2*P3*P4*P5*P6*P7*P8 = 0{,}96*0{,}89*0{,}86*0{,}64*0{,}89*0{,}96*0{,}89*0{,}86 = 0{,}31$$

As is seen from the computation the lowest failure-free operation is in:

- Synchronizing sleeve
- Gear shift fork
- Gearwheels meshed

Let us increase the above units' reliability to 95% and calculate:

$$P = P1*P2*P3*P4*P5*P6*P7*P8 = 0{,}96*0{,}89*0{,}95*0{,}95*0{,}89*0{,}96*0{,}89*0{,}95 = 0{,}56$$

System updating to the required level

Units' reliability rate must be redistributed for the whole system reliability to be $P^{TP} = 0.47$
Solution:
The whole system reliability is
$P = 0.31$

To update the system let us consider the units in the reliability ascending sequence

$$\begin{matrix} P_4 & P_3 & P_8 & P_2 & P_5 & P_7 & P_1 & P_6 \\ 0.64< & 0.86< & 0.86< & 0.89< & 0.89< & 0.89< & 0.96< & 0.96 \\ i_1 & i_2 & i_3 & i_4 & i_5 & i_6 & i_7 & i_8 \end{matrix}$$

Assign the relevant numbers.
Value the rank:

$$r_1 = \left(\frac{P^{TP}}{P_2*P_3*P_4*P_5*P_6*P_7*P_8*P_9}\right)^{\frac{1}{1}} = \left(\frac{0.47}{0.86*0.86*0.89*0.89*0.89*0.96*0.96*1}\right)^{\frac{1}{1}} = 0.98 \quad (2)$$

$$r_2 = \left(\frac{P^{TP}}{P_3*P_4*P_5*P_6*P_7*P_8*P_9}\right)^{\frac{1}{2}} = \left(\frac{0.47}{0.86*0.89*0.89*0.89*0.96*0.96*1}\right)^{\frac{1}{2}} = 0.92 \quad (3)$$

$$r_3 = \left(\frac{P^{TP}}{P_4*P_5*P_6*P_7*P_8*P_9}\right)^{\frac{1}{3}} = \left(\frac{0.47}{0.89*0.89*0.89*0.96*0.96*1}\right)^{\frac{1}{3}} = 0.9 \quad (4)$$

$$r_4 = \left(\frac{P^{TP}}{P_5*P_6*P_7*P_8*P_9}\right)^{\frac{1}{4}} = \left(\frac{0.47}{0.89*0.89*0.96*0.96*1}\right)^{\frac{1}{4}} = 0.89 \quad (5)$$

$$r_5 = \left(\frac{P^{TP}}{P_6*P_7*P_8*P_9}\right)^{\frac{1}{5}} = \left(\frac{0.47}{0.89*0.96*0.96*1}\right)^{\frac{1}{5}} = 0.89 \quad (6)$$

$$r_6 = \left(\frac{P^{TP}}{P_7*P_8*P_9}\right)^{\frac{1}{6}} = \left(\frac{0.47}{0.96*0.96*1}\right)^{\frac{1}{6}} = 0.894 \quad (7)$$

$$r_7 = \left(\frac{P^{TP}}{P_8*P_9}\right)^{\frac{1}{7}} = \left(\frac{0.47}{0.96*1}\right)^{\frac{1}{7}} = 0.9 \quad (8)$$

$$r_8 = \left(\frac{P^{TP}}{P_9}\right)^{\frac{1}{8}} = \left(\frac{0.47}{1}\right)^{\frac{1}{8}} = 0.91 \quad (9)$$

Let us estimate the "K" number (Table 2)

Table 2 – Calculation of updated units

№	P_i	r_i	Symbol
1	0.64	0.98	<
2	0.86	0.92	<

3	0.86	0.9	<
4	0.89	0.894	<
5	0.89	0.894	<
6	0.89	0.894	<
7	0.96	0.9	>
8	0.96	0.91	>

To update the system to goal the reliability of «K=6» units must be raised. Their reliability is raised to the level $P_n^{TP} = r_k = r_6 = 0.894$.

The system reliability is estimated:
$$P=\left(P_0^{TP}\right)^K * \prod_{i=k+1}^{n+1} P_i = \left(P_0^{TP}\right)^6 * P_7 * P_8 * P_9 = (0.894)^6 * 0.96 * 0.96 * 1 = 0.47$$

To update the system the following must be done: the reliability of the first unit is 0.96; the reliability of the second unit must be raised from 0.89 to 0.894; the reliability of the third unit must be increased from 0.86 to 0.894; the reliability of the fourth unit must be raised from 0.64 to 0.894; the reliability of the fifth unit must be raised from 0.89 to 0.894; the reliability of the sixth unit is 0.96; the reliability of the seventh unit must be increased from 0.89 to 0.894; the reliability of the eighth unit must be raised from 0.86 to 0.894.

Review

The reliability probability of the rest car SU is $P_Д = 0.85 - 0.9$. The reliability probability based on experimental data and after updating the system to the required level is P = 0.31, $P_{mp\cdot}$ = 0.47. In comparing the obtained results of reliability probability with the acceptable value margin we can see that the obtained results do not enter the range. It cannot be stated that SU of the gear box must be updated to the acceptable reliability level as the failure data are insufficient. It makes it possible to conclude that failure data are not enough.

Conclusion

In this research an analysis of "GASon Next" gear box was made. Based on it, the reliability structural diagram was made. The obtained data enable us to make a histogram of average trouble-free life which shows the main units subject to risk. In calculating gear box SU reliability the permissible reliability level is lower than failure-free operation probability of the other car SU which do not respond to traffic safety. One can conclude that it is necessary to increase reliability of the gear box units, particularly synchronizing sleeves, gear shift forks and gearwheels meshed.

Acknowledgements

This research done with the financial support from Ministry of Education and Science of the Russian Federation in the frame of the complex project "The establishment of the high-tech manufacturing of safe and export-oriented GAZ vehicles with autonomous control systems and the possibility of integration with the electric platform on the base of components of Russian production" under the contract №03.G25.31.0270 from 29.05.2017 (Governmental Regulation №218 from 09.04.2010).The experimental research was conducted with the use of measurement equipment of the NNSTU Centre of collective using "Transport Systems".

Bibliography

[1] Kuzmin N.A., Borisov G.V. "Car working capacity change laws" – NNSTU, - N. Novgorod, 2014.
[2] Berdnikov L.A., Suvorov I.A., Vilkov N.A. Comprehensive approach to examining the factors which determine the reliability of turbochargers of car ICE. Collected Papers of 12 International Young People's Scientific and Technical Conference "The Future of Technical Science", Nizhny Novgorod, NNSTU, - 2013. Pp. 195-196.
[3] Korchazhkin M.G., Khoryayev E.A. Research of fuel system reliability of car diesel engines. Collected Papers of the 94 International Scientific and Technical Conference of the Association of car engineers "Non-piloted vehicles: problems and prospects". Nizhny Novgorod, NNSTU – 2016, pp. 240-245.
[4] Korchazhkin A.V., Solovyev S.S. Functional reliability features of car automatic transmission. Papers of NNSTU after R.Y. Alexeev – 2013, № 4 (101). Nizhny Novgorod, 2013, pp. 66-71.

[5] Tokarev A.V., Berdnikov L.A. Diagnostic specificity of brake systems in checkup. Transport Systems – 2017, №3. Nizhny Novgorod, 2017, pp. 21-26.
[6] Korchazhkin M.G. Increasing functional reliability of city bus engines when operating in a high heat load mode / M.G. Korchazhkin // Thesis for a candidate of technical sciences, 2005.
[7] Kustikov A.D. City bus transmission reliability problems / A.D. Kustikov, N.A. Kuzmin, M.G. Korchazhkin // Papers of NNSTU after R.Y. Alexeev – 2013, №4 (101). Pp. 18-26.
[8] L. A. Berdnikov, A. A. Pikulkin, M.G. Korchazhkin, P. I. Bazhan and L. A. Zakharov. Study of factors affecting the reliability of turbochargers/ IOP Conference Series: Materials Science -and Engineering, Volume 386, conference 1

IOP Conf. Series: Materials Science and Engineering **695** (2019) 012018 doi:10.1088/1757-899X/695/1/012018

Designed brake system reliability assessment techniques application on the basis of the real one

R.R. Yusupov, A.V. Alexeev, A.D. Yarzhemsky, M.I. Chichkina, L.A. Berdnikov

Nizhny Novgorod State Technical University after R.Y. Alexeev

E-mail: ro.iusupov@yandex.ru

Abstract. The paper analyses and assesses the reliability of the designed brake system on the basis of the real one. Analysis of the basic diagram and the components' eventual failures was made. Based on the built structural diagram of the brake system the reliability level was estimated. Measures to increase the system reliability were proposed.

Key words: brake system, reliability calculation, failure quota, operating time.
Research subject: GAZel NEXT hydraulic brake system reliability calculation.
Research purpose: build a structural diagram and estimate the level of reliability of the hydraulic brake system, refine to the required reliability level.

Introduction

Nowadays, the majority of people have a car at their disposal. Using any vehicle is permissible if it has a faultless brake system. The ability of a forced deceleration and a quick stop is the automobile most important characteristic which influences its performance criteria (efficiency, fuel consumption, and other factors); they are essential for traffic safety. The brake system technical state affects traffic safety sufficiently.

The car brake system means, firstly, its safety, therefore, it must be paid great attention to, it must be timely maintained and correctly used. In case the brake system fails in the vehicle operation the consequences may be grave both for the driver and the people involved.

Nowadays, new safety measures are being devised to prevent road accidents caused by the brake system, so reliability is the main quality indicator.

"GAZel NEXT" hydraulic brake system reliability calculation.

Reliability calculation plan.

1. The hydraulic brake system analysis and investigation of possible failures of the system components.
2. Building a system reliability structural diagram.
3. Mean time to failure analysis.
4. System reliability computing.
5. System updating to the required level.

Hydraulic brake system diagram analysis

To analyze the above the general view of the hydraulic brake system is shown below (fig. 1).

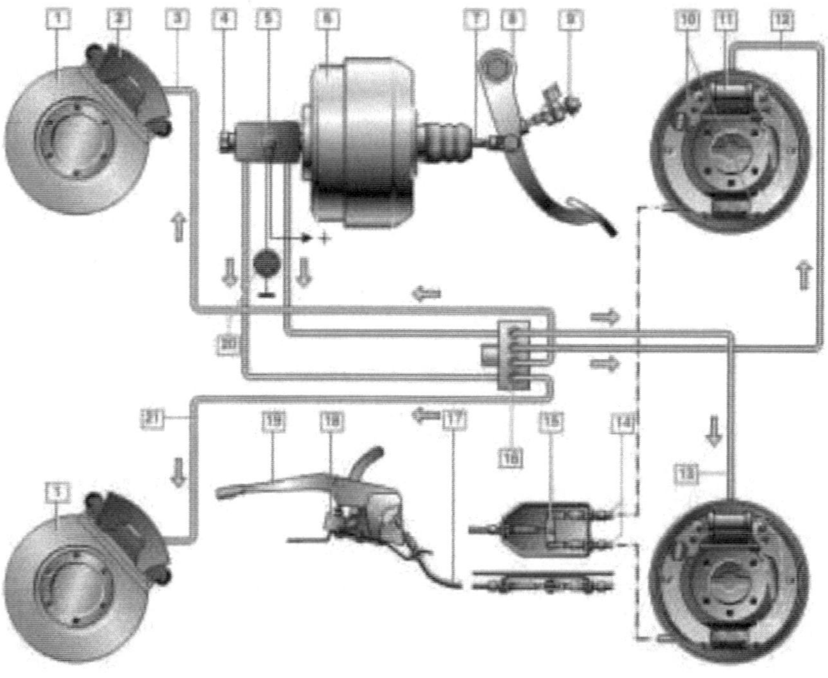

Figure 1. GAZel NEXT break system diagram

1 – Brake disk; 2 – Front wheel brake gear; 3 – Right front wheel brake gear brake pipe; 4 – Brake master cylinder; 5 – Hydraulic drive tank; 6 – Vacuum booster; 7 – Tappet; 8 – Brake pedal; 9 – Brake pedal sensor and stop light switch; 10 – Rear brake gear shoes; 11 – Wheel brake cylinder; 12 - Right rear wheel brake gear brake pipe; 13 – Left rear wheel brake gear brake pipe; 14 – Parking brake rear cables; 15 – Equalizer; 16 – ABS unit; 17 – Front cable of the parking brake; 18 – Parking brake signaling device switch; 19 – Brake handle; 20 – Brake fluid alert signaling device; 21 – Front left wheel brake gear break pipe.

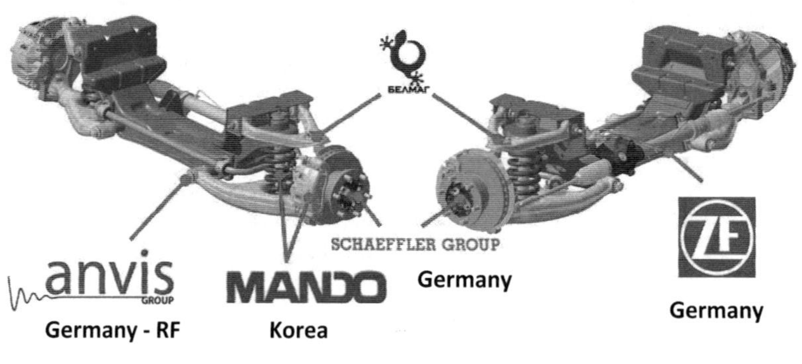

Figure 2. GAZel NEXT brake system components' manufacturers

Building system reliability structural diagram

The building is done including the following assumption:

1) All failures are independent;
2) In case one component fails the system remains non-serviceable.

- Brake pedal
- Brake pedal sensor and stop light switch
- Tappet

- Vacuum booster
- Hydraulic drive tank
- Brake master cylinder
- ABS unit
- Front axle brake gear
- Rear axle brake gear
- Brake fluid alert level signaling device

Let us build a system reliability model of the obtained components (fig. 3):

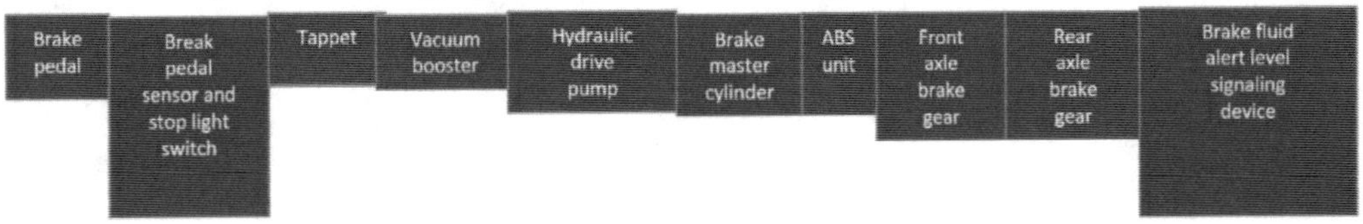

Figure 3. Hydraulic brake system reliability diagram

Mean time to failure analysis

The initial data array was obtained from the of dealer stations OAO GAZ information on the mean time to failure (table. 1)

Table 1 - Failure data

Components	Number of failures	Mean time to failure, km	Reconditioning mean time, norm/h
Brake pedal	-	$\to\infty$	$\to 0$
Brake pedal sensor and stop light switch	-	$\to\infty$	$\to 0$
Tappet	-	$\to\infty$	$\to 0$
Vacuum booster	1	27	1,4
Hydraulic drive tank	-	$\to\infty$	$\to 0$
Brake master cylinder	-	$\to\infty$	$\to 0$
ABS unit	-	$\to\infty$	$\to 0$
Front axle brake gear	4	60160	0,78
Rear axle brake gear	23	79360	0,445
Brake fluid alert level signaling device	1	15048	0,4

Total number of failures : 29

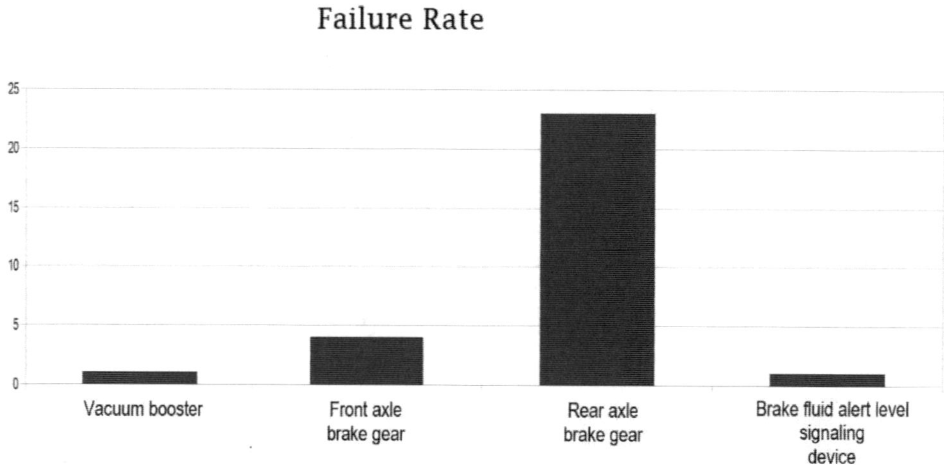

Figure 4. System components' failure distribution

System reliability calculation

The system reliability is estimated:

$$P = P_1 * P_2 * P_3 * P_4 * P_5 * P_6 * P_7 * P_8 * P_9 * P_{10} \quad (1)$$

The probability of no-failure operation according to experimental data about the system failures is the following:

Brake pedal $P_1 \geq 0{,}98$
Brake pedal sensor and stop light switch $P_2 \geq 0{,}98$
Tappet $P_3 \geq 0{,}98$
Vacuum booster $P_4 \geq 0{,}97$
Hydraulic drive tank $P_5 \geq 0{,}98$
Brake master cylinder $P_6 \geq 0{,}98$
ABS unit $P_7 \geq 0{,}98$
Front axle brake gear $P_8 \geq 0{,}86$
Rear axle brake gear $P_9 \geq 0{,}21$
Brake fluid alert signaling device $P_{10} \geq 0{,}97$

P=0,98*0,98*0,98*0,97*0,98*0,98*0,98*0,86*0,21*0,97=0,15

As is obvious the component Rear axle brake gear has low probability of no-failure operation. Let us assume that we raise its reliability to 95% and do the calculation.

P=0,98*0,98*0,98*0,97*0,98*0,98*0,98*0,86*0,95*0,97=0,68

The system updating to the required reliability level

To update the system, the components' reliability norms redistribution is required so that the whole system reliability must be $P^{TP} = 0.71$, as the allowable failure rate of the systems responsible for traffic safety is (2-10 %).

Solution:

- Whole system reliability
- Brake pedal P_1
- Brake pedal sensor and stop-light switch P_2
- Tappet P_3
- Vacuum booster P_4
- Hydraulic drive tank P_5
- Brake master cylinder P_6

- ABS unit P_7
- Front axle brakes gear P_8
- Rear axle brakes gear P_9
- Brake fluid alert level signaling device P_{10}
- To update the system let us consider the components in reliability ascending sequence

$$\begin{array}{cccccccccc} P_9 & P_8 & P_4 & P_{10} & P_2 & P_3 & P_5 & P_6 & P_7 & P_1 \\ 0.21< & 0.86< & 0.97< & 0.97< & 0.98< & 0.98< & 0.98< & 0.98< & 0.98< & 0.98 \\ i_1 & i_2 & i_3 & i_4 & i_5 & i_6 & i_7 & i_8 & i_9 & i_{10} \end{array}$$

where i – is the order number

Let us assign respective numbers.
Introduce $P_{11} = 1$ which is an imaginary element of the system

- Let us value ranks:

$$r_1 = \left(\frac{P^{TP}}{P_2 * P_3 * P_4 * P_5 * P_6 * P_7 * P_8 * P_9}\right)^{\frac{1}{1}} = \left(\frac{0.71}{0{,}86 * 0.97 * 0.97 * 0.98 * 0.98 * 0.98 * 0.98 * 0.98 * 0.98 * 1}\right)^{\frac{1}{1}} = 0.99 \quad (2)$$

$$r_2 = \left(\frac{P^{TP}}{P_2 * P_3 * P_4 * P_5 * P_6 * P_7 * P_8 * P_9}\right)^{\frac{1}{2}} = \left(\frac{0.71}{0.97 * 0.97 * 0.98 * 0.98 * 0.98 * 0.98 * 0.98 * 0.98 * 1}\right)^{\frac{1}{2}} = 0.92 \quad (3)$$

$$r_3 = \left(\frac{P^{TP}}{P_3 * P_4 * P_5 * P_6 * P_7 * P_8 * P_9}\right)^{\frac{1}{3}} = \left(\frac{0.71}{0.97 * 0.98 * 0.98 * 0.98 * 0.98 * 0.98 * 0.98 * 1}\right)^{\frac{1}{3}} = 0.94 \quad (4)$$

$$r_4 = \left(\frac{P^{TP}}{P_4 * P_5 * P_6 * P_7 * P_8 * P_9}\right)^{\frac{1}{4}} = \left(\frac{0.71}{0.98 * 0.98 * 0.98 * 0.98 * 0.98 * 0.98 * 1}\right)^{\frac{1}{4}} = 0.95 \quad (5)$$

$$r_5 = \left(\frac{P^{TP}}{P_5 * P_6 * P_7 * P_8 * P_9}\right)^{\frac{1}{5}} = \left(\frac{0.71}{0.98 * 0.98 * 0.98 * 0.98 * 0.98 * 1}\right)^{\frac{1}{5}} = 0.95 \quad (6)$$

$$r_6 = \left(\frac{P^{TP}}{P_6 * P_7 * P_8 * P_9}\right)^{\frac{1}{6}} = \left(\frac{0.71}{0.98 * 0.98 * 0.98 * 0.98 * 1}\right)^{\frac{1}{6}} = 0.96 \quad (7)$$

$$r_7 = \left(\frac{P^{TP}}{P_7 * P_8 * P_9}\right)^{\frac{1}{7}} = \left(\frac{0.71}{0.98 * 0.98 * 0.98 * 1}\right)^{\frac{1}{7}} = 0.96 \quad (8)$$

$$r_8 = \left(\frac{P^{TP}}{P_8 * P_9}\right)^{\frac{1}{8}} = \left(\frac{0.71}{0.98 * 0.98 * 1}\right)^{\frac{1}{8}} = 0.96 \quad (9)$$

$$r_9 = \left(\frac{P^{TP}}{P_9}\right)^{\frac{1}{9}} = \left(\frac{0.71}{0.98 * 1}\right)^{\frac{1}{9}} = 0.96 \quad (10)$$

$$r_{10} = \left(\frac{P^{TP}}{P_9}\right)^{\frac{1}{10}} = \left(\frac{0.71}{1}\right)^{\frac{1}{10}} = 0.97 \quad (11)$$

Let us estimate the «K» number which is a reliability element (table 2)

Table 2 - Number of updated components calculation

№	P_i	r_i	Symbol
1	0.21	0.99	<
2	0.86	0.92	<
3	0.97	0.94	>
4	0.97	0.95	>
5	0.98	0.95	>
6	0.98	0.96	>
7	0.98	0.96	>
8	0.98	0.96	>
9	0,98	0,96	>
10	0,98	0,97	>

- To update the system to the required level the reliability of «K=2» components must be increased. Their reliability is raised to the level: $P_n^{\text{TP}} = r_k = r_2 = 0.92$.

The system reliability is calculated as follows:

$$P=(P_0^{\text{TP}})^K * \prod_{i=k+1}^{n+1} P_i = (P_0^{\text{TP}})^2 * P_3 * P_4 = (0.92)^2 * 0.97 * 0{,}97 * 0{,}98 * 0{,}98 * 0{,}98 * 0{,}98 * 0{,}98 * 0{,}98 * 0{,}98 * 1 = 0.71 \quad (12)$$

To update the system the following must be done: the reliability of the first component is 0.98; the reliability of the second component is 0,98; the reliability of the third component is 0.98; the reliability of the fourth component is 0.97; the reliability of the fifth component is 0,98; the reliability of the sixth component is 0,98; the reliability of the seventh component is 0.98; the reliability of the eighth component must be raised from 0,86 to 0,92; the reliability of the ninth component must be raised from 0,21 to 0,92; the reliability of the tenth component is 0,97.

Reliability probability according to the observed data of the hydraulic brake system is P = 0.15 < P_{add}. As the reliability of the rear axle brake gear is lower than the allowable one. Reliability level of the rear axle brake gear must be raised to 0.92.

The system updating to goal was done, reliability level must be raised from P = 0.15 to P = 0.71.

Conclusion

In the course of investigation, the hydraulic brake system was analyzed, based on it a reliability structural diagram was built. Proceeding from the obtained data a mean time to failure histogram was built; it shows the main components subject to risk. In calculating the system reliability it was found that to raise the whole system reliability to the required level the brake gear reliability of front and rear axles must be increased to 0,92.

Acknowledgements

This research done with the financial support from Ministry of Education and Science of the Russian Federation in the frame of the complex project "The establishment of the high-tech manufacturing of safe and export-oriented GAZ vehicles with autonomous control systems and the possibility of integration with the electric platform on the base of components of Russian production" under the contract №03.G25.31.0270 from 29.05.2017 (Governmental Regulation №218 from 09.04.2010).The experimental research was conducted with the use of measurement equipment of the NNSTU Centre of collective using "Transport Systems".

Bibliography

[1] Kuzmin N.A., Borisov G.V. "Car working capacity change laws" – NNSTU, - N. Novgorod, 2014.
[2] Berdnikov L.A., Suvorov I.A., Vilkov N.A. Comprehensive approach to examining the factors which determine the reliability of turbochargers of car ICE. Collected Papers of 12 International Young People's Scientific and Technical Conference "The Future of Technical Science", Nizhny Novgorod, NNSTU, - 2013. Pp. 195-196.
[3] Korchazhkin M.G., Khoryayev E.A. Research of fuel system reliability of car diesel engines. Collected Papers of the 94 International Scientific and Technical Conference of the Association of car engineers "Non-piloted vehicles: problems and prospects". Nizhny Novgorod, NNSTU – 2016, pp. 240-245.
[4] Korchazhkin A.V., Solovyev S.S. Functional reliability features of car automatic transmission. Papers of NNSTU after R.Y. Alexeev – 2013, № 4 (101). Nizhny Novgorod, 2013, pp. 66-71.
[5] Tokarev A.V., Berdnikov L.A. Diagnostic specificity of brake systems in checkup. Transport Systems – 2017, №3. Nizhny Novgorod, 2017, pp. 21 26.

[6] Korchazhkin M.G. Increasing functional reliability of city bus engines when operating in a high heat load mode / M.G. Korchazhkin // Thesis for a candidate of technical sciences, 2005.
[7] Kustikov A.D. City bus transmission reliability problems / A.D. Kustikov, N.A. Kuzmin, M.G. Korchazhkin // Papers of NNSTU after R.Y. Alexeev – 2013, №4 (101). Pp. 18-26.
[8] L A Berdnikov, A A Pikulkin, M G Korchazhkin, P I Bazhan and L A Zakharov Study of factors affecting the reliability of turbochargers/ IOP Conference Series: Materials Science and Engineering, Volume 386, conference 1

IOP Conf. Series: Materials Science and Engineering **695** (2019) 012019 doi:10.1088/1757-899X/695/1/012019

Application of assessment techniques of reliability of the designed steering system based on the real one

R.R. Yusupov, A.V. Alexeev, A.D. Yarzhemsky, M.G. Korchazhkin, A.S. Brovkin

Nizhny Novgorod State Technical University after R.Y. Alexeev

E-mail: ro.iusupov@yandex.ru

Abstract. The paper analyses and assesses the designed steering system reliability based on the real one. An analysis of the basic diagram and components' eventual failures was made. Based on the built structural diagram of the steering the reliability level was estimated. Measures to enhance the system reliability were suggested.

Key words: steering system, reliability calculation, failure rate, running hours.
Study: "GAZel NEXT" hydraulic steering reliability calculation.
Study purpose: build a structural diagram and estimate the hydraulic steering system reliability level, update the reliability to goal.

Introduction

Steering is a number of units and mechanisms aimed at synchronizing the steering wheel position (steering) and the car wheels' turning angle. The basic function of steering in any vehicle is to provide the angle and keep the direction preset by the driver.

Steering is one of the most important systems of any car, the systems that provide traffic safety. Prescriptions of traffic rules are to prove it; vehicle operation with the specified non-serviceable mechanisms is strictly forbidden, therefore, the steering system reliability must be at the required level.

"GAZel NEXT" hydraulic steering reliability calculation.

Reliability calculation plan.

1. Hydraulic steering and the system possible failure analysis.
2. Building a structural diagram of the system reliability.
3. Mean time to failure analysis.
4. System reliability calculation.
5. System updating to the required reliability level.

Hydraulic steering diagram analysis

The general view of the hydraulic steering is shown for the analysis (Fig. 1).

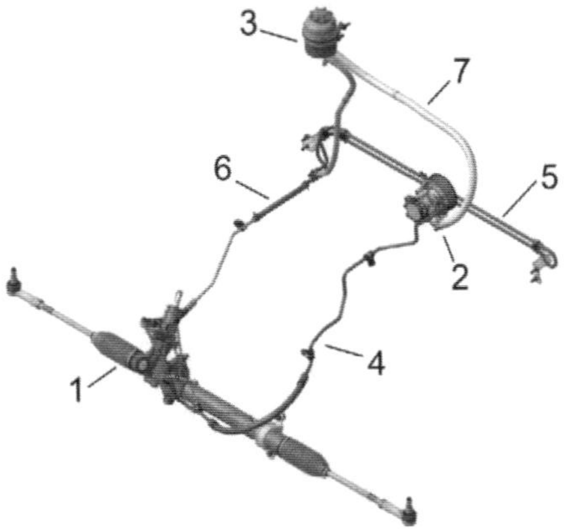

Figure 1. GAZel NEXT steering diagram
(1 - integrated steering; 2 – steering booster pump; 3 - tank; 4 – pressure hose; 5 – radiator; 6 – drain line; 7 – suction hose)

Building structural diagram of the system reliability
Building is made taking into account the following assumptions:
1) All the failures are independent;
2) In case of a component's failure the system becomes non-serviceable.

- Steering wheel
- Steering shaft
- Integrated steering
- Steering booster pump
- Steering booster radiator
- Steering booster tank
- Main line
- Control arm
- Steering wheel end

Let us make the system reliability model with the obtained components (Fig. 2):

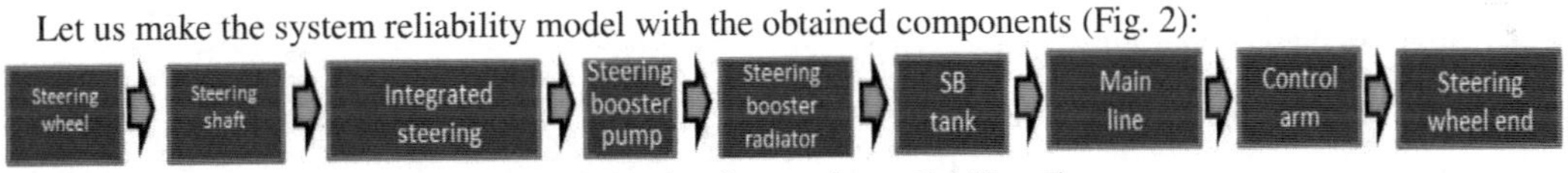

Figure 2. Hydraulic steering reliability diagram

Mean time to failure analysis
The initial data array was obtained from the OAO GAZ dealer stations' information as to the failure mean time (Table 1)

Table 1 - Failure data

Components	Number of failures	Failure mean time, km	Recovery mean time, norm/hour
Steering wheel	3	167023	0.42
Steering shaft	12	49198	0,7
Integrated steering	1	84	3,5
Steering booster pump	7	54342	0,697
Steering booster radiator	-	$\to\infty$	$\to 0$
Steering booster tank	2	71585	0,47
Main line	1	18	0,44
Control arm	1	141347	0,36
Steering wheel end	1	29909	0,36

Total number of failures: 28

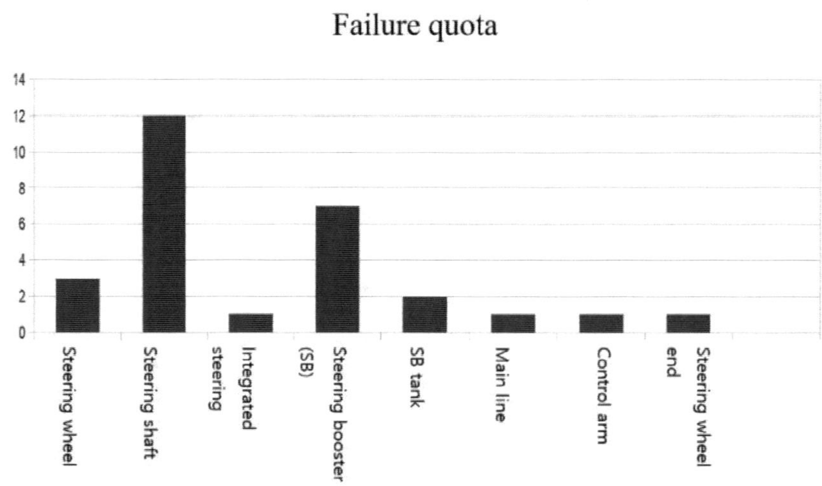

Figure 3. System components failure distribution

System reliability calculation

The system reliability is estimated:

$$P = P_1 * P_2 * P_3 * P_4 * P_5 * P_6 * P_7 * P_8 * P_9 \quad (1)$$

Probability of no-failure based on the experimental data about the system failures is:

Steering wheel $P_1 \geq 0.89$
Steering shaft $P_2 \geq 0.57$
Integrated steering $P_3 \geq 0.96$
Steering booster (SB) pump $P_4 \geq 0.75$
SB radiator $P_5 \geq 0.98$
SB tank $P_6 \geq 0.92$
Main line $P_7 \geq 0.96$
Control arm $P_8 \geq 0.96$
Steering wheel end $P_9 \geq 0.96$

P=0.89*0.57*0.96*0.75*0.98*0.92*0.96*0.96*0.96=0.29

The component Steering wheel has a low probability of no-failure work. To increase the system reliability the component's reliability must be raised to 95%. The system calculated value is:

P=0.89*0.95*0.96*0.75*0.98*0.92*0.96*0.96*0.96=0.49

System update to goal

To update the system the components' reliability rates must be redistributed so that the whole system reliability equals $P^{TP} = 0.5$, as the allowable value of failures in the systems to meet traffic safety requirements is (2-10 %)

Solution:

- The total system reliability

Steering wheel P_1
Steering shaft P_2
Integrated steering P_3
SB pump P_4
SB radiator P_5
SB tank P_6
Main line P_7
Control arm P_8
Steering wheel end P_9

To update the system the components are considered in reliability ascending sequence being assigned the due numbers

$$\begin{matrix} P_2 & P_4 & P_1 & P_6 & P_3 & P_7 & P_8 & P_9 & P_5 \\ 0.57< & 0.75< & 0.89< & 0.92< & 0.96< & 0.96< & 0.96< & 0.96< & 0.98 \\ i_1 & i_2 & i_3 & i_4 & i_5 & i_6 & i_7 & i_8 & i_9 \end{matrix}$$

where i – is the order number

The following value is introduced $P_{10} = 1$ as an imaginary component of the system

- Ranks are valued:

$$r_1 = \left(\frac{P^{TP}}{P_2*P_3*P_4*P_5*P_6*P_7*P_8*P_9*P_{10}}\right)^{\frac{1}{1}} = \left(\frac{0.5}{0.75*0.89*0.92*0.96*0.96*0.96*0.96*0.98*1}\right)^{\frac{1}{1}} = 0.98 \quad (2)$$

где P^{TP} – требуемая надежность системы

$$r_2 = \left(\frac{P^{TP}}{P_3*P_4*P_5*P_6*P_7*P_8*P_9*P_{10}}\right)^{\frac{1}{2}} = \left(\frac{0.5}{0.89*0.92*0.96*0.96*0.96*0.96*0.98*1}\right)^{\frac{1}{2}} = 0.86 \quad (3)$$

$$r_3 = \left(\frac{P^{TP}}{P_4*P_5*P_6*P_7*P_8*P_9*P_{10}}\right)^{\frac{1}{3}} = \left(\frac{0.5}{0.92*0.96*0.96*0.96*0.96*0.98*1}\right)^{\frac{1}{3}} = 0.87 \quad (4)$$

$$r_4 = \left(\frac{P^{TP}}{P_5*P_6*P_7*P_8*P_9*P_{10}}\right)^{\frac{1}{4}} = \left(\frac{0.5}{0.96*0.96*0.96*0.96*0.98*1}\right)^{\frac{1}{4}} = 0.88 \quad (5)$$

$$r_5 = \left(\frac{P^{TP}}{P_6*P_7*P_8*P_9*P_{10}}\right)^{\frac{1}{5}} = \left(\frac{0.5}{0.96*0.96*0.96*0.98*1}\right)^{\frac{1}{5}} = 0.9 \quad (6)$$

$$r_6 = \left(\frac{P^{TP}}{P_7*P_8*P_9*P_{10}}\right)^{\frac{1}{6}} = \left(\frac{0.5}{0.96*0.96*0.98*1}\right)^{\frac{1}{6}} = 0.91 \quad (7)$$

$$r_7 = \left(\frac{P^{TP}}{P_8*P_9*P_{10}}\right)^{\frac{1}{7}} = \left(\frac{0.5}{0.96*0.98*1}\right)^{\frac{1}{7}} = 0.91 \quad (8)$$

$$r_8 = \left(\frac{P^{TP}}{P_9*P_{10}}\right)^{\frac{1}{8}} = \left(\frac{0.5}{0.98*1}\right)^{\frac{1}{8}} = 0.92 \quad (9)$$

$$r_9 = \left(\frac{P^{TP}}{P_{10}}\right)^{\frac{1}{9}} = \left(\frac{0.5}{1}\right)^{\frac{1}{9}} = 0.93 \quad (10)$$

Ratio of «K» components is estimated, the components whose reliability enhances (Table. 2)

Table 2 - The number of updated components estimation

№	P_i	r_i	Symbol
1	0.57	0.98	<
2	0.75	0.86	<
3	0.89	0.87	>
4	0.92	0.88	>
5	0.96	0.9	>
6	0.96	0.91	>
7	0.96	0.91	>
8	0.96	0.92	>
9	0,98	0,93	>

- To update the system to goal the reliability of «K=2» components must be increased. Their reliability is raised to the level $P_n^{TP} = r_k = r_2 = 0.86$.

The system reliability is estimated:

$$P=\left(P_0^{TP}\right)^K * \prod_{i=k+1}^{n+1} P_i = \left(P_0^{TP}\right)^2 * P_3 * P_4 * P_5 * P_6 * P_7 * P_8 * P_9 * P_{10} = \\ =(0.86)^2 * 0.89 * 0{,}92 * 0{,}96 * 0{,}96 * 0{,}96 * 0{,}96 * 0{,}98 * 1 = 0.5 \quad (11)$$

To update the system the following must be done: the reliability of the first component is 0.89; the reliability of the second component must be raised from 0.57 to 0.86; the reliability of the third component is 0.96; the reliability of the fourth component must be increased from 0,75 to 0.87; the reliability of the fifth component is 0,98; the reliability of the sixth component is 0,92; the reliability of the seventh component is 0.96; the reliability of the eighth component is 0,96; the reliability of the ninth component is 0,96.

The no-failure operation probability based on the experimental data of the hydraulic steering system is P = 0.29 < P_{add}. As the steering shaft reliability is lower than acceptable reliability. The reliability level of the steering shaft must be raised to 0.95.

The system update to the required level of reliability is made; the reliability level must be raised from P = 0.29 to P = 0.5.

Conclusion

An analysis of the hydraulic steering system was made in the research. Hence, the structural reliability diagram was made. The obtained data made it possible to construct a mean time to failure histogram; it shows the main components subject to risk. In calculating the system reliability it was found that to increase the whole system reliability to goal the reliability of the steering shaft and of the SB pump must be raised to 0,86 and 0,87 respectively.

Acknowledgements

This research done with the financial support from Ministry of Education and Science of the Russian Federation in the frame of the complex project "The establishment of the high-tech manufacturing of safe and export-oriented GAZ vehicles with autonomous control systems and the possibility of integration with the electric platform on the base of components of Russian production" under the contract №03.G25.31.0270 from 29.05.2017 (Governmental Regulation №218 from 09.04.2010).The experimental research was conducted with the use of measurement equipment of the NNSTU Centre of collective using "Transport Systems".

Bibliography

[1] Kuzmin N.A., Borisov G.V. "Car working capacity change laws" – NNSTU, - N. Novgorod, 2014.

[2] Berdnikov L.A., Suvorov I.A., Vilkov N.A. Comprehensive approach to examining the factors which determine the reliability of turbochargers of car ICE. Collected Papers of 12 International Young People's Scientific and Technical Conference "The Future of Technical Science", Nizhny Novgorod, NNSTU, - 2013, pp. 195-196.

[3] Korchazhkin M.G., Khoryayev E.A. Research of fuel system reliability of car diesel engines. Collected Papers of the 94 International Scientific and Technical Conference of the Association of car engineers "Non-piloted vehicles: problems and prospects". Nizhny Novgorod, NNSTU – 2016, pp.240-245.

[4] Korchazhkin A.V., Solovyev S.S. Functional reliability features of car automatic transmission. Papers of NNSTU after R.Y. Alexeev – 2013, № 4 (101). Nizhny Novgorod, 2013, pp. 66-71

[5] Tokarev A.V., Berdnikov L.A. Diagnostic specificity of brake systems in checkup. Transport Systems – 2017, №3. Nizhny Novgorod, 2017, pp. 21-26.

[6] Korchazhkin M.G. Increasing functional reliability of city bus engines when operating in a high heat load mode / M.G. Korchazhkin // Thesis for a candidate of technical sciences, 2005.

[7] Kustikov A.D. City bus transmission reliability problems / A.D. Kustikov, N.A. Kuzmin, M.G. Korchazhkin // Papers of NNSTU after R.Y. Alexeev – 2013, №4 (101). Pp. 18-26.

[8] L A Berdnikov, A A Pikulkin, M G Korchazhkin, P I Bazhan and L A Zakharov Study of factors affecting the reliability of turbochargers/ IOP Conference Series: Materials Science and Engineering, Volume 386, conference 1

Bearing floatation of forest machines (theoretical calculation)

E G Khitrov [1], A V Andronov [2]

[1]Associate Professor, Department of Technological Processes and Machines of the Forestry Complex, St. Petersburg State Forestry University named after S. M. Kirov, St. Petersburg, Russia
[2]Associate Professor, Department of Forestry Engineering, Service and Repair, St. Petersburg State Forestry University named after S. M. Kirov, St. Petersburg, Russia

E-mail: andronovalexandr@gmail.com

Abstract. The purpose of the article is to summarize the calculation procedure and calculate the forest machines floatation depending on the type of mover and soil conditions. Research methods are analysis of literary sources, numerical methods of applied mathematics during calculations and calculated data approximation. Formulas and the procedure for calculating the track depth, the coefficients of resistance to movement, adhesion and traction of the forest machines movers are given. In the compiled mathematical model, slipping is not an input parameter; its value is determined by calculation comparing the resistance forces to movement and the mover adhesion to the bearing surface. According to the calculations results, it was found that forestry machines with a caterpillar mover have greater cross-country ability compared to wheeled vehicles. On weak and moderate soils, the bearing floatation loss of tracked vehicles due to insufficient mover adhesion to the bearing surface does not occur. A factor limiting the pressure of the caterpillar mover on the bearing surface should be recognized as the track depth, since at a pressure of more than 0.04 MPa, the track depth exceeds the machine clearance. To ensure the bearing floatation of wheeled vehicles on weak soils, the pressure of the wheeled mover should be limited to 0.6 MPa. To ensure the bearing floatation of wheeled vehicles on moderate soils, the pressure of the wheeled mover should be limited to 0.95 MPa.

Introduction

The bearing floatation of forest machines is usually estimated by the traction coefficient, which is the difference between the adhesion coefficient and the resistance coefficient to movement [1], [2]. A positive value of the traction coefficient means that the machine is able to move along the bearing surface, a negative value indicates the loss of bearing pass ability [1], [2]. A theoretical assessment of the thrust, adhesion, and drag coefficients is the most complicated task for forest engineering [2] - [5]. To date, based on the methods of road transport theory in off-road conditions, studies have been carried out to model the interaction of the forest machines movers with various soils and forest soil [1], [2], [6] - [10]. As a result, engineering dependencies were obtained for assessing track formation, traction and adhesive properties, and equipment floatation in predetermined soil conditions. At the same time, the authors note [1], [2], [6] - [10] that further research is required for the further development and improvement of the presented mathematical models. In particular, in the well-known works [1], [2], [6] - [10], the skidding coefficient of the mover is considered an input parameter of the models, which is not entirely correct, since the skidding coefficient value is a derivative that depends on the interaction parameters of the mover with deformable soil [11], [12].

The purpose of the article is to summarize the calculation procedure and perform the calculation of the forest vehicles floatation depending on the type of mover and soil conditions.

Research methods are analysis of literary sources, numerical methods of applied mathematics in calculations and computational data approximation.

Problem setting. As it was noted in the introduction, the basic machine floatation will be evaluated using the traction coefficient φ_p [11], [12]:

$$\varphi_p = \varphi_r - \mu , \tag{1}$$

where φ_p is the movement resistance coefficient, μ is the adhesion coefficient.

The movement resistance coefficient is found by the formula [11], [12]:

$$\varphi_r = \frac{F_r}{G_w}, \tag{2}$$

where F_r is the resistance force to movement, consisting of the deformable soil resistance, the sticky soil resistance to the mover separation, the mover deformation resistance, G_w is the reduced load on a single mover.

At this stage of the study only the deformable soil resistance is taken into account.

The mover adhesion coefficient to the support surface is calculated by the formula [11], [12]:

$$\mu = \frac{F_T}{G_w}, \tag{3}$$

where F_T is the mover adhesion force to the support surface.

The force with which the soil resists the mover movement is defined as the integral [11], [12]:

$$F_r = B\int_0^h p\,dh, \tag{4}$$

where p is the mover normal pressure on the soil (the average on the contact spot), B is the contact spot width (assuming a rectangular spot shape), h is the track depth (soil sediment).

Here are the formulas for calculating the track depth (sediment). It is based on the equation for the upsetting of a stamp pressed into a deformable half-space [10]:

$$h = \frac{JpaBp_s}{(p_s - p)\sqrt{E(E - Jp)}}\,\text{arctg}\left(\frac{E(H-h)}{aB\sqrt{E(E-Jp)}}\right), \tag{5}$$

where J is a coefficient of accounting for the rectangular contact spot aspect ratio, a is a power factor of the deformable soil layer, H is the thickness of the deformable soil layer, E is the modulus of the total soil deformation, p_s is soil bearing capacity.

The coefficients J, a are calculated by the formulas [1], [11], [12]:

$$J = \frac{0{,}03 + \frac{L}{B}}{0{,}6 + 0{,}43\frac{L}{B}}, \tag{6}$$

$$a = 0{,}64\frac{B+H}{H}, \tag{7}$$

where L is the mover contact spot length with the ground.

For a crawler, the contact spot length is taken as the input parameter of the model. For a wheeled tractor, the contact spot length is approximately determined by the formula [13]:

$$L = \frac{d}{2}, \tag{8}$$

where d is the wheel diameter, for a more accurate calculation of the contact spot length, the joint deformations of the elastic mover and the ground should be considered.

To calculate the load-bearing capacity, the refined formulas are proposed [14]:

$$\begin{cases} p_s = p_{s0}\alpha_Z \\ p_{s0} = 0{,}5K_1N_1\gamma B + N_2\gamma h + K_3N_3C \\ K_1 = \dfrac{L}{L+0{,}4B};\ K_3 = \dfrac{L+B}{L+0{,}5B} \\ N_1 = \dfrac{1-T^4}{T^5};\ N_2 = \dfrac{1}{T^2};\ N_3 = \dfrac{2(1+T^2)}{T^3};\ T = \text{tg}\left(\dfrac{\pi}{4} - \dfrac{\varphi}{2}\right) \\ \alpha_Z = 1 + \dfrac{H^*h}{2H\cdot(H-h-0{,}25H^*)};\ H^* = \dfrac{\sqrt{2}}{2}\exp\left[\left(\dfrac{\pi}{4}+\dfrac{3\varphi}{4}\right)\tan\dfrac{3\varphi}{4}\right]B\cos\dfrac{3\varphi}{4}tg\varphi \end{cases} \tag{9}$$

where p_{s0} is soil bearing capacity with unlimited capacity of the deformable layer, α_Z is the power factor of the deformable soil layer, K_1, K_3 are the ratio coefficients of the mover's contact spot length and width with the bearting surface, N_1, N_2, N_3 are the coefficients of the soil internal friction angle, C is the specific soil adhesion, φ is the soil internal friction angle, γ is the specific soil gravity, T, H * are the auxiliary notations.

The mover adhesion force to the ground is found as an integral [1], [11], [12]:

$$F_T = B\int_0^L \tau dx\,, \tag{10}$$

where x is the horizontal coordinate measured along the contact spot length, τ is the tangential shear stress distributed over the contact spot.

The tangential stress is calculated by the equation [1], [11], [12]:

$$\tau = \frac{jG\tau_{\max}}{jG + t_g\tau_{\max}}\,, \tag{11}$$

where j is the soil shear strain, G is the soil shear modulus, tg is the lug pitch, and τ_{max} is the maximum soil shear resistance.

The shear deformation is found by the formula [1], [2]:

$$j = Sx\,, \tag{12}$$

where S is the slip coefficient.

The maximum resistance to the soil shear is found by the formula [1], [11], [12]:

$$\tau_{\max} = p\,\mathrm{tg}\,\varphi + C\left(1 - \frac{j}{t_g}\right), \tag{13}$$

The dependence integration (10), taking into account expressions (11) – (13), results in the formula:

$$F_T = \frac{BG^2L(p\,\mathrm{tg}\,\varphi + C)}{(G-C)^2} - S\cdot\frac{BCGL^2}{2t_g(G-C)} - \frac{1}{S}\cdot\frac{(p\,\mathrm{tg}\,\varphi + C)^2BG^2t_g}{(G-C)^3}\ln\left(1 + S\frac{L(G-C)}{t_g(p\,\mathrm{tg}\,\varphi + C)}\right) \tag{14}$$

The input parameters of the mathematical model characterizing the soil are: the total strain modulus E, the shear modulus G, the specific adhesion C, the internal friction angle φ, the specific gravity γ, the power (thickness) of the deformable layer H. The mover parameters: the reduced load G_w or the average pressure on soil p, the contact width B and the contact spot length L (depending on the mover's geometrical parameters), the lug pitch t_g.

When the initial data are received, the average pressure range p is divided into intervals with a given pitch Δp. The system of equations (5) - (9) is solved numerically for each pressure value p. As a result, a table of the correspondence between the average pressure p and the track depth h is obtained. Using the table numerically (e.g., using the trapezoids method) the approximate value of the integral formula is found (4), then the resistance factor φ_r according to the formula (2) is calculated. For each pressure value p two slipping coefficient values S are defined. The first value is found as the solution of equation (14) at $F_T = F_r$ (slipping is sufficient to ensure adhesion). Then, for each p value, the function maximum (14) and the corresponding value of the slipping coefficient S are numerically determined (at this coefficient value, the mover adhesion with the support surface is maximum). The adhesion coefficient is found by the formula (3). The traction coefficient is calculated by the formula (1).

Calculation results and conclusions

For the calculation in this study the following initial data is taken: 1) wheel width B = 0,7 m, wheel diameter d = 1,6 m, lug pitch t_g = 0,14 m, average mover pressure p = 0,1 ... 0,1 MPa, pressure pitch Δp = 0,1 MPa; 2) caterpillar width B = 0,6 m, the caterpillar contact spot length with bearing surface L = 6 m, lug pitch t_g = 0,15 m, average mover pressure p = 0,01 ... 0,1 MPa, pressure pitch Δp = 0.01 MPa. The forest soil properties are taken in accordance with table 1.

Table 1. Forest soil properties [10]

Parameter	Soil conditions category	
	Weak soil, category III	Moderate soil, category II
γ, MN / m^3	0,0075	0,0085
E, MPa	0,4	1
C, MPa	0,0053	0,0108
G, MPa	0,22	0,25
φ, °	12	14
H, m	0,8	0,4

The calculation results are presented in figures 1-3.

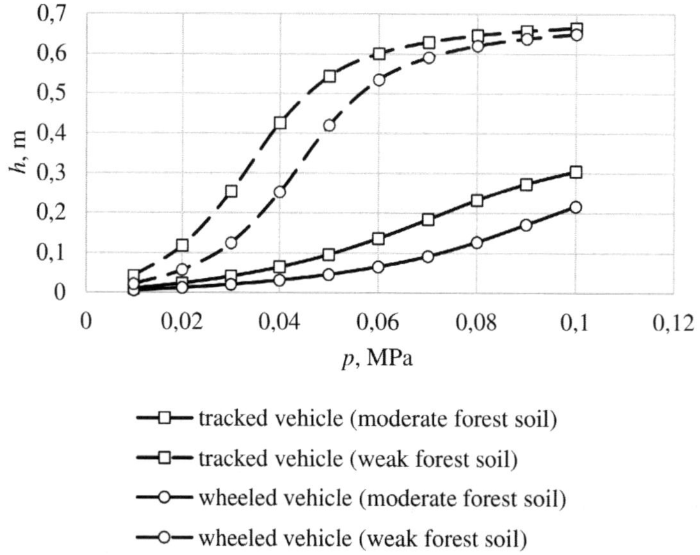

Figure 1. Track depth (sediment) under the influence of a wheeled and caterpillar mover

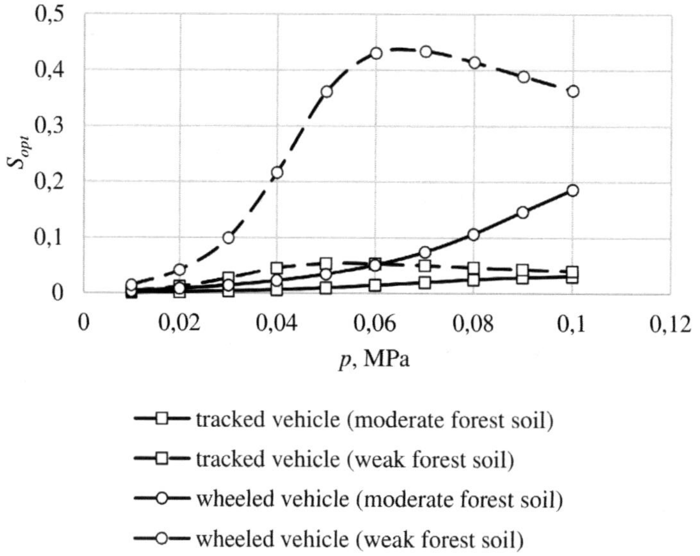

Figure 2. Slipping coefficient at maximum traction of the wheeled and caterpillar mover with the supporting surface

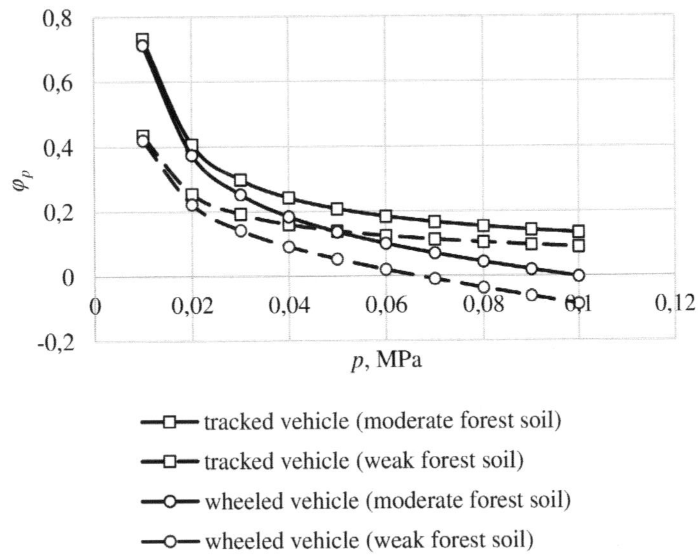

Figure 3. Traction coefficient of wheeled and caterpillar mover

Research results

By the calculation results it is established that:

1. Forest vehicles with a caterpillar engine have a greater floatation compared to wheeled vehicles. For the weak soils and the moderate soils the loss of the bearing terrain tracked vehicles because of the poor mover adhesion with the bearing surface does not occur. The factor limiting the caterpillar mover pressure on the bearing surface should be recognized as the depth of the track, since at a pressure of more than 0.04 MPa, the track depth exceeds the machine clearance.

2. To ensure the wheeled vehicles on the weak soils, the wheeled mover pressure should be limited to 0.6 MPa. To ensure the wheeled vehicles basic floatation on the moderate soils, the wheeled mover pressure should be limited to 0.95 MPa.

In the compiled mathematical model, slipping is not an input parameter; its value is determined by calculation based on the results of comparing the movement resistance forces and the mover adhesion with the bearing surface. This builds on the results obtained in previous studies on the forest vehicles floatation. The mathematical model uses a simplified expression to determine the contact spot length; in the long term it is expedient to clarify the calculation procedure when taking into account the influence of the wheel drive elasticity on the contact spot shape.

References:

[1] Lukhminsky V.A. Improving models and methods for predicting the tracked forest machines floatation. Diss. Cand. tech. sciences. Place of protection: Arkhangelsk, NArFU 2018 p 179.

[2] Khakhina A.M. Methods for predicting and increasing the wheeled forest machines floatation. Diss. Doct. tech. sciences. Place of protection: Arkhangelsk, NArFU 2018 p 318.

[3] Anisimov G.M. The main directions of the logging industry scientific and technological progress Bulletin of the St. Petersburg Forestry Academy. SPb . FTA. No. 169. 2003 pp. 129-140.

[4] Anisimov G.M. Forecasting methods of the logging machines technical level. SPb. Izvestia SPb FTA. 2005 pp. 4-10.

[5] Anisimov G.M., Bolshakov B.M. New concepts of the logging machines theory. SPb FTA, 1998 p 114.

[6] Bozhbov V.E. Improving the skidding process efficiency by justifying the forwarders' workload. Abstract of diss. Cand. tech. sciences. Arkhangelsk: NArFU, 2015 p 20.

[7] Dmitrieva M.N. Modeling the interaction of a wheeled mover of small-sized forestry machines with weak bearing soil. Abstract of diss. Cand. tech. sciences. Arkhangelsk: NArFU 2018 p 20.

[8] Kalistratov A.V. Modeling of cyclic compaction in the problems of reducing the negative impact of forest machines on the soil. Abstract of diss. Cand. tech. sciences. Arkhangelsk: NArFU 2016 p 20.

[9] Peskov V.B. Improving models for assessing soil rutting and compaction under the wheeled forest machines movers' influence. Abstract of diss. Cand. tech. sciences. Arkhangelsk: NArFU, 2018 p 20.
[10] Kochnev A., Khitrov E. Theoretical models for rut depth evaluation after a forestry machine's wheel Passover. International Multidisciplinary Scientific GeoConference Surveying Geology and Mining Ecology Management, SGEM 18. 2018. pp. 1005-1012.
[11] Ageikin Ya.S. All-terrain wheeled and combined movers. Theory and calculation. M. Mechanical Engineering 1972 p 184.
[12] Ageikin Ya.S. Cars passability. M. Mechanical Engineering 1981 p 232.
[13] Saarilahti M. «Development of a protocol for ecoefficient wood harvesting on sensitive sites (Ecowood). Soil interaction model». University of Helsinki, Department of Forest Resource Management, 2002. 39 p.
[14] Larin V.V. Prediction methods of multi-axle wheeled vehicles bearing floatation on the ground: Diss. Doct. tech. Sciences: 05.05.03.-M., 2007 p 530.

IOP Conf. Series: Materials Science and Engineering **695** (2019) 012021 doi:10.1088/1757-899X/695/1/012021

Comparison of the forest machine mover average and nominal pressure on the soil

E G Khitrov [1], A V Andronov [2]
[1]Associate Professor, Department of Technological Processes and Forestry Machines, Kirov St. Petersburg State Forestry University, St. Petersburg, Russia
[2]Associate Professor, Department of Forest Engineering, Service and Repair, St. Petersburg State Forestry University named after S.M. Kirov, St. Petersburg, Russia

E-mail: andronovalexandr@gmail.com

Abstract. The purpose of this article is to investigate the relationship between the estimated values of nominal and average wheeled mover pressure on the ground and a rigid bearing surface. Research methods are mathematical analysis, computational experiment, and calculated data approximation. The article presents and compares methods of calculating average mover pressure in deformable soils, medium mover pressure on a rigid base, and nominal pressure on soil. According to the calculations results, it was found that the average mover pressure on firm soils almost corresponds to the estimate of the mover pressure on a rigid base. The nominal pressure is lower than the pressure on the rigid base by 40-50 kPa. The average pressure estimate on moderate soils is close to the nominal pressure. The average pressure on weak soils in the load range of 10-20 kN is close to the nominal pressure, at high values of wheel load the average pressure on the soils category is estimated to be twice less than the nominal pressure of the mover on the soils. It has been established by calculation that the average pressure of the wheeled mover on the deformable forest soils and on a rigid base is connected with the value of the mover nominal pressure on the soils, used by the foreign researchers. The dependences are linear and include correction factors, the values of which are determined by the mover parameters (wheel width and diameter, tyre pressure, and wheel load) and the forest soil category. The received equations allow simplifying the interaction estimates comparison process of forest machine movers on the soils, given within the domestic and foreign techniques limits.

Introduction

Contact spot pressure is one of the most important indicators that characterizes the machine's engine impact on the bearing surface (soils). There are various methods and models for calculating the track depth, soil compaction and logging equipment floatation, based on the comparison of the mover pressure and the soil bearing capacity [1], [2]. However, the authors use different approaches to determine the pressure: in foreign studies the nominal ground pressure (NGP) [3] - [7] is more frequent, in domestic studies the specific or average pressure of the mover on the contact spot p and on the rigid surface q [8], [9]. Based on the NGP value, the expected maximum pressure of the MMP (mean maximum pressure) [4], [5] is determined, and the maximum mover pressure on the ground p_{max} q is calculated using correction coefficients at p, q values [8], [9]. MMP, p_{max} values are also used to calculate the machine's interaction with soils. Moreover, in the scientific literature and periodicals there is not enough information on how to compare the values of NGP, p, q and, therefore, the recommendations of various authors.

The purpose of this article is to investigate the relationship between the design values of nominal and average wheeled mover pressure on the soils and a rigid support surface.

Research methods are mathematical analysis, computational experiment, and calculated data approximation.

Setting the objective and main results of the study

The equation for the average pressure on the wheeled mover rectangular spot with the deformable soils is written down as follows:

$$p = \frac{G_w}{bl}, \tag{1}$$

where G_w is the wheel load, b is the contact spot width, l is the contact spot length.

The contact spot width is found by the equation [10]:

$$b = B + \frac{10h \cdot h_Z}{1 - h + H_T - h_Z}, \tag{2}$$

where B is the tyre width, H_T is the tyre height, h_Z is the tyre radial deformation, h is the track depth.

The contact spot length is found by the equation [10]:

$$l = \sqrt{dh_Z - h_Z^2} + \sqrt{d \cdot (h_Z + h) - (h_Z + h)^2}, \tag{3}$$

where d is the wheel diameter.

The track depth is found from the solution of the differential equation of pressing the stamp into the deformable soils, which, taking into account the compression deformation and the array shifting, has the form [11]:

$$h = \frac{abCp_S Jp}{\sqrt{p_S(-Cp_S Jp + p_S - Jp)}} \operatorname{arctg}\left(\frac{p_S}{ab\sqrt{p_S(-Cp_S Jp + p_S - Jp)}} (H - h) \right) \tag{4}$$

where a is a parameter that takes into account the deformable soils layer thickness, J is a parameter that takes into account the contact spot length and width ratio, H is the deformable soils layer thickness, p_S is the soils bearing capacity, C is the soils stiffness, determined depending on its rheological model.

The conclusion of the equation (4) and the rationale for its structure are detailed in the paper [11].

The parameters a, J are calculated according to the known formulas [10]:

$$a = 0{,}64 \frac{b + H}{H}, \tag{5}$$

$$J = \frac{0{,}03b + l}{0{,}43b + 0{,}6l}, \tag{6}$$

It was shown [11] that the Maxwell Thompson's rheological body model with equal modules of instantaneous and long-term deformation correctly describes the behavior of the forest soils under the machine load. The expression for the rigidity of such a model is [12]:

$$C = \frac{2}{E} - \frac{1}{E} \exp\left(-\frac{E}{\eta} t \right), \tag{7}$$

where E is the deformation modulus, n is the soils viscosity, t is the exposure time.

The exposure time is determined by the formula:

$$t = n \frac{l}{v}, \tag{8}$$

where n is the number of wheels passing successively along the track, v is the machine speed.

The equations analysis (1) - (8) shows that the values of 12 variables are needed to determine the average pressure p: G_w, B, H_T, d, p_w, v, n, E, η, p_S, h, h_z. In this case, h is found in equation (4) by the iteration method, but an unknown quantity h_z is present in the equation.

The empirical equation allows calculating the elastic wheel pressure on the ground at a known radial deformation h_z [10]:

$$p = \frac{p_w + q}{2} \cdot \frac{\pi h_Z}{B} \cdot \left(\frac{B}{H_T} + \frac{3H_T}{B} \right) \cdot \left(1 - \frac{h_Z}{B} \right), \tag{9}$$

where p_w is the pressure in the tyre.

The pressure on the tyre contact spot with the rigid bearing surface can be expressed as follows:

$$q = \frac{G_w}{b_0 l_0}, \tag{10}$$

where b_0, l_0 is the width and length of the wheel contact spot with a rigid surface according to formulas (2), (3) at $h = 0$:

$$b_0 = B, \tag{11}$$

$$l_0 = 2\sqrt{dh_{Z0} - h_{Z0}^2}, \tag{12}$$

where h_{z0} is the wheel radial deformation the on a rigid surface, determined by the Heideckel formula [10]:

$$h_{Z0} = \frac{G_w}{\pi p_w \sqrt{Bd}}, \tag{13}$$

The right parts of formulas (1) and (9) are equated; then, taking into account (2), (3), (10) – (13), an equation with the variables G_w, B, d, p_w, h and h_Z is obtained. The values G_w, B, d, p_w refer to the original data, then, at their given values, the equation will contain two unknown quantities, therefore, on its basis an expression for h_Z depending on h is obtained.

The calculations analysis with the initial data variation shows that the calculated values of h_Z are almost exactly expressed in terms of h with the help of the fourth degree polynomials:

$$h_Z = a_0 + a_1 h + a_2 h^2 + a_3 h^3 + a_4 h^4, \tag{14}$$

where a_0, a_1, a_2, a_3, a_4 are the approximation coefficients determined by the initial data G_w, B, d, p_w.

The calculation procedure is as follows: after specifying the initial data (G_w, B, H_T, d, p_w, v, n, E, η, p_S,) the equations (1) - (9) are solved taking into account (2), (3), (10) - (13) as a system with respect to h_Z when varying h in a given range, the calculated data is approximated and the function $h_Z = f(h)$ as an expression (14) is obtained. Then, by iteration, the equation (4) with respect to h taking into account the formulas (1) – (3), (5) – (8), (14) is solved.

The influence of the parameters G_w, B, H_T, d, p_w, v, E, η, p_S on the pressure p in more detail is investigated. To do this, a computational experiment in the *Maple* 2017 environment is applied; the ranges of parameter-factors values changes are presented in table 1.

Table 1 - Range of changes in factors in the computational experiment to study the influence of the mover and soil parameters on the average pressure on the contact spot

Parameter	Dimension	Minimum value	Maximum value
d	m	1,2	1,8
L	m	0.5	0.8
H_T	m	0.25	0.65
P_W	MPa	0.15	0.55
G_w	kN	10	70
v	m/s	0.25	1.5
H	m	0,2	1
E	MPa	0.4	3
η	MPa s	1	10
p_S	MPa	0.05	0.55

The calculations were performed with 1000 factor values combinations generated as random variables evenly distributed in the ranges according to table 1 ($\approx 2^N$ combinations). According to the processing results the calculated data using the least squares method, excluding insignificant factors, p [kPa] for the pressure is obtained:

$$p = 37{,}52 d^{-0{,}38} B^{-0{,}41} p_w^{0{,}14} G_w^{0{,}45} E^{0{,}16} p_S^{0{,}35}, \tag{15}$$

The dimensions of the equation parameters (15) correspond to those specified in table 1.

So it can be concluded that the mover contact spot pressure with soils is determined by such factors as the wheel diameter d, the tyre width B, the internal tyre pressure p_w, the wheel load G_w, the deformation modulus E and the bearing soils capacity p_S.

To calculate the nominal pressure *NGP* the formula is used [4], [5]:

$$NGP = \frac{2G_w}{Bd}, \tag{16}$$

IOP Conf. Series: Materials Science and Engineering **695** (2019) 012021 doi:10.1088/1757-899X/695/1/012021

The compared results of calculations by formulas (10), (15), (16) for soils of three categories (I - firm E = 3 MPa, p_S = 0.3 MPa; II - moderate, E = 1 MPa, p_S = 0.1 MPa, III - weak, E = 0.4 MPa, p_S = 0.04 MPa) at B = 0.7 m, d = 1.6 m, p_w = 0.35 MPa are presented in table 2.

Table 2 – Calculation results of nominal and average mover pressure on soils

G_w, кN	NGP, kPa	q, kPa	p, kPa		
			Soils category		
			weak (III)	moderate (II)	firm (I)
10	18	61	25	39	69
20	36	87	34	54	94
30	54	106	41	65	113
40	71	123	46	74	129
50	89	138	51	81	143
60	107	152	55	88	155
70	125	164	59	95	166

The tabular data are illustrated by the graphs in figure 1.

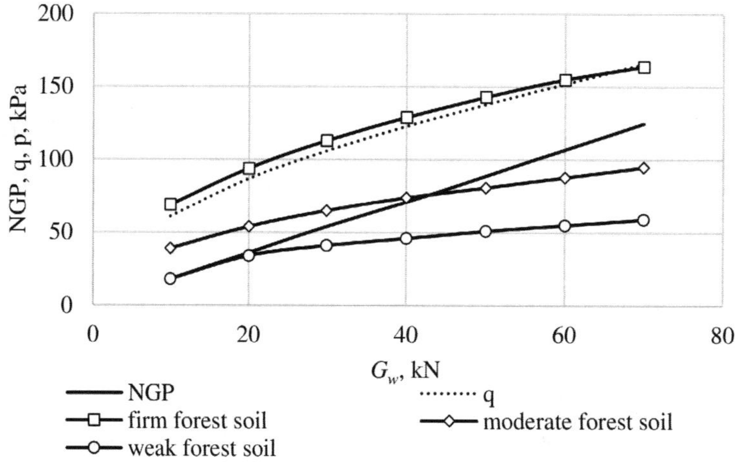

Figure 1. Nominal and average mover pressure on the soils, depending on the load

The analysis of the calculation results shows that the average mover pressure on the soils category I p practically corresponds to the mover pressure estimate on the rigid base q. The nominal NGP pressure is lower than the pressure on the rigid base q by 40-50 kPa. The average pressure assessment on the soils category II is close to the nominal pressure NGP. The average pressure on soils category III in the range of 10-20 kN load is close to the nominal pressure of the NGP. For the large load values on the wheel the average pressure on soils category III is roughly twice less than the nominal mover pressure on the soils, which can be explained by increasing the mover contact area with weak soils.

Processing the results of the calculations using equations (10), (15), (16), with variation of the initial data in the range specified in Table 1, shows that the average pressure on the deformable soil p, pressure on a rigid base q and the nominal pressure of NGP can be correlated with high accuracy by simplified dependencies (R2>0.95):

$$p = K_p \psi_p NGP \, , \tag{17}$$

$$q = K_q \psi_q NGP \, , \tag{18}$$

$$p = K \psi q \, , \tag{19}$$

Where ψ_p, ψ_q, ψ are auxiliary coefficients depending on the mover parameters:

$$\psi_p = B^{\frac{1}{2}} d^{\frac{1}{2}} G_w^{-\frac{1}{2}} \, , \tag{20}$$

$$\psi_q = B^{\frac{1}{2}} d^{\frac{1}{2}} p_w^{\frac{1}{2}} G_w^{-\frac{1}{2}}, \quad (21)$$

$$\psi = B^{\frac{1}{3}} p_w^{-\frac{1}{3}}, \quad (22)$$

where *Kq, Kp, K* are numerical coefficients depending on the soils category, the values of which are presented in table 3.

Table 3. Numerical coefficients for calculating the average pressure on the soils on the nominal pressure

Coefficient	Soil category		
	weak (III)	moderate (II)	firm (I)
K_p	3,8	6,0	10,5
K_q	17,1	17,1	17,1
K	0,30	0,47	0,83

In formulas (20) – (22), the quantities dimensions correspond to those specified in table 1.

Conclusion

It was found by calculation that the wheeled mover average pressure on the deformable forest soils and on the rigid base is associated with the nominal mover pressure value on the soils, used by foreign researchers. The dependences are linear and include correction factors, the values of which are determined by the mover parameters (wheel width and diameter, tyre pressure, wheel load) and the forest soils category.

The received equations will allow simplifying the estimates comparison process of forest machine movers interaction on the soil, given within the limits of domestic and foreign techniques.

Referances:

[1] Khakhina A.M. Methods for predicting and increasing the floatation of wheeled forest machines. Diss. Doct. tech. sciences. Place of protection: Arkhangelsk,NArFU 2018 p 318.
[2] Khitrov E.G., Khakhina A.M., Dmitrieva M.N., Peskov VB, Grigoryeva O.I. An updated model for evaluating the traction properties of a forest machine wheeled mover. Proceedings of the St. Petersburg Forestry Academy. 2016 No p 217.
[3] Bekker M.G. Introduction to Terrain-Vehicle Systems. The University of Michigan Press: Ann Arbor, 1969.
[4] Saarilahti M. Development of a protocol for ecoefficient wood harvesting on sensitive sites (Ecowood). Dynamic terrain classification». University of Helsinki, Department of Forest Resource Management, 2002. 22 p.
[5] Saarilahti M. Development of a protocol for ecoefficient wood harvesting on sensitive sites (Ecowood). Soil interaction model». University of Helsinki, Department of Forest Resource Management, 2002. 39 p.
[6] Wong J.Y. Terramechanics and Off-Road Vehicle Engineering: Terrain Behaviour Off-Road Vehicle Performance and Design, Elsevier second edition, 2010.
[7] Wong J.Y. Theory of Ground Vehicles. John Wiley and Sons, New Jersey, fourth edition, 2008.
[8] Anisimov G.M., Bolshakov B.M. New concepts of the logging machines theory. St. Petersburg: FTA, 1998 114 p.
[9] Anisimov G.M., Bolshakov B.M. Fundamentals of minimizing soil compaction with skidding systems. St. Petersburg FTA, 1998 106p.
[10] Ageikin Ya.S. Passability of cars. M. Mechanical Engineering, 1981 232 p.
[11] Peskov V.B. Improving models for assessing soil's rutting and compaction under the influence of wheeled forest machines movers. Diss. Cand. tech. sciences. Place of protection: Arkhangelsk, NArFU 2018 p 190.
[12] Blend D. Theory of linear viscoelasticity. Moscow, World 1965 p 200.

IOP Conf. Series: Materials Science and Engineering **695** (2019) 012022 doi:10.1088/1757-899X/695/1/012022

Engine-and-transmission installation frequency analysis of a wheeled forestry tractor

Churakov A.V.[1], Sokolova V.A.[1], Markov V.A.[1], Parfenopulo G.K.[1], Kalyashov V.A.[2]
[1]St. Petersburg State Forestry University named after S.M. Kirov, Institutsky per., 5, St. Petersburg, Russian Federation
[2]St. Petersburg State University of Architecture and Civil Engineering, 2nd Krasnoarmeyskaya St., 4, 190005, St. Petersburg

E-mail: sheff_2.01@mail.ru

Received xxxxxx
Accepted for publication xxxxxx
Published xxxxxx

Abstract: The article provides a frequency analysis of an engine-and-transmission installation of a wheeled forestry tractor. It can be used for machine-building enterprises of the forest complex. During the wheeled forestry tractors operation in the engine-and-transmission installation, resonance phenomena may occur when the internal frequencies of the system coincide with the external disturbance. The solution of the problem was carried out on the basis of the developed equivalent dynamic circuit, which was described by the differential equations of torsional vibration. As a result of solving these equations, the values of the engine-and-transmission installation natural frequencies were obtained. A comparative analysis of the frequencies obtained with regard to the damping properties of the system has been made, and possible resonance zones have been identified. The information obtained allows selecting the optimal damping indicators for the wheeled forestry tractor engine-and-transmission installation operation in a resonant-free mode.
Keywords: engine-and-transmission installation, resonance, torsional vibrations

Introduction.

Changes in the transmission parameters of a wheeled forestry tractor and the introduction of new technical solutions affect the inertial and stiff, damping properties of the transmission, and, consequently, its eigen frequencies values. As the researchers note, the dynamic component of the tangential traction force is the result of self-oscillations, has a certain dominant frequency, which can cause resonant oscillatory processes of the "engine-transmission-tractor with a bundle" system. The oscillatory system eigen frequencies can vary over a wide range when installing elements with a high compliance coefficient. In connection with the change in the transmission parameters or the new technical solutions introduction, for example, damping elements, it is necessary to conduct a frequency analysis of the oscillatory system and develop measures that exclude system's resonant modes.

The studies analysis of transport and technological machines [1, 2, 3, 4, 6] shows that the resonant modes of the lugs stimulus with the bearing surface may arise at the third natural frequency of the dynamic system "engine - transmission - tractor with a bundle" natural oscillations. Therefore, an equivalent dynamic system design scheme must have at least four discrete masses bonded together by inertia-free connections, and the suspensions stiffness of transmission elements must also be taken into account. In order to analyze the simplifications of the design scheme, an analytical determination of the natural transmission frequencies values was made, taking into account the reactive circuits for the seven mass systems. Figure 1 shows the design scheme, and their reduced parameters are indicated.

The basis for obtaining the differential equations of the wheeled forestry tractor transmission torsional oscillations is the Langrangian equations of the second kind:

$$\frac{d}{dt}\left(\frac{\partial T}{\partial \dot{q}}\right)-\frac{\partial T}{\partial q}+\frac{\partial \Pi}{\partial q}+\frac{\partial \Phi}{\partial \dot{q}}=Q_i(t) \quad (1)$$

where T is the kinetic energy of the system; P is potential energy of the system; Φ is dissipative function; q_i is generalized coordinates; Q_i is perturbing generalized forces.

For the generalized coordinates q_i the angles φ_i of masses rotation fixed on the shafts are taken. They are counted from the system equilibrium deformed position, assuming that these angles in the equilibrium position are equal to zero. As generalized forces in this case, the torques M_{kp} applied to the corresponding elements of the system are used.

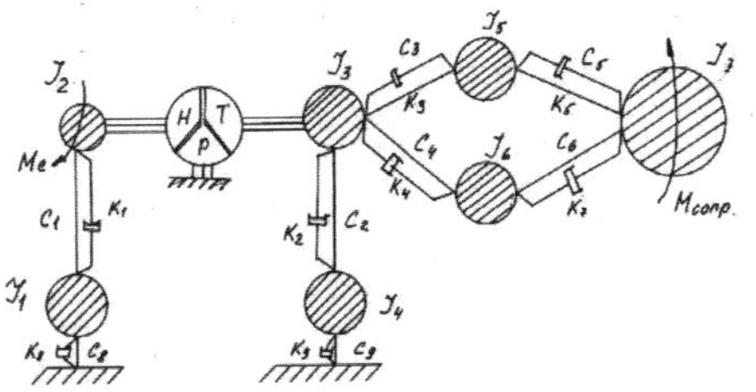

Figure 1. Estimated equivalent 7 mass scheme of a dynamic system "engine – transmission - tractor with a bundle", taking into account reactive circuits:

J_1, J_4 is the inertia moment of the engine and gearbox, respectively, at turning in the transverse plane relatively to the center of gravity under the influence of reactive moment;

J_2, J_3, J_5, J_6, is the given moments of the transmission sections inertia;

J_7 is inertia moment of the tractor and bundle translational moving mass;

C_1, C_2 is the given to the angular stiffness of the suspension engine and transmission;

C_3-C_7 is the crankshaft stiffness coefficient of the transmission sections reduced to the crankshaft axis;

C_8, C_9 is the reduced to the angular stiffness of the tractor mass in the transverse plane;

k_1 ... k_9 is the damping coefficients of the corresponding transmission sections.

Substituting the found values of T and P into the corresponding Langrange equations and performing the corresponding transformations, a system of differential equations describing the transmission shafts torsional vibrations is obtained

$$\begin{gathered}
J_1\ddot{\varphi}_1+k_2(\dot{\varphi}_1-\varphi_2)+c_2(\varphi_1-\varphi_2)-k_1\varphi_1-c_1\varphi_1=0\,;\\
J_2\ddot{\varphi}_2+k_3(\dot{\varphi}_2-\dot{\varphi}_3)+c_3(\varphi_2-\varphi_3)-k_2(\dot{\varphi}_1-\dot{\varphi}_2)-c_2(\varphi_1-\varphi_2)=M_c\,;\\
J_3\ddot{\varphi}_3+A_1\dot{\varphi}_3+A_2\dot{\varphi}_5-k_3(\dot{\varphi}_2-\dot{\varphi}_3)-c_3(\varphi_2-\varphi_3)=0\\
J_4\ddot{\varphi}_4+k_5\dot{\varphi}_4+c_5\varphi_4-k_4(\dot{\varphi}_5-\dot{\varphi}_4)-c_4(\varphi_5-\varphi_4)=\dot{\varphi}_3(A_5-A_1)+\dot{\varphi}_5(A_4-A_2)\\
J_6\ddot{\varphi}_6+k_8(\dot{\varphi}_6-\dot{\varphi}_8)+c_8(\varphi_6-\varphi_8)-k_6(\dot{\varphi}_5-\dot{\varphi}_6)-c_6(\varphi_5-\varphi_6)=0;\\
J_7\ddot{\varphi}_7+k_9(\dot{\varphi}_7-\dot{\varphi}_9)+c_9(\varphi_7-\varphi_9)-k_7(\dot{\varphi}_5-\dot{\varphi}_7)-c_7(\varphi_5-\varphi_7)=0;\\
J_8\ddot{\varphi}_8+k_{10}(\dot{\varphi}_8-\dot{\varphi}_{10})+c_{10}(\varphi_8-\varphi_{10})-k_8(\dot{\varphi}_6-\dot{\varphi}_8)+c_8(\varphi_6-\varphi_8)=0;\\
J_9\ddot{\varphi}_9+k_{11}(\dot{\varphi}_9-\dot{\varphi}_{10})+c_{11}(\varphi_9-\varphi_{10})-k_9(\dot{\varphi}_7-\dot{\varphi}_9)-c_9(\varphi_8-\varphi_9)=0;\\
J_{10}\ddot{\varphi}_{10}+k_{10}(\dot{\varphi}_8-\dot{\varphi}_{10})+c_{10}(\varphi_8-\varphi_{10})-k_{11}(\dot{\varphi}_9-\dot{\varphi}_{10})-c_{11}(\varphi_9-\varphi_{10})=-M_{cnp}.
\end{gathered} \quad (2)$$

Differential equations of any system forced oscillations with the presence of disturbing moments applied to the elements

$$Q_i(t) = Q_{i0}\,Cos\,pt, \quad (3)$$

where Q_{i0} is the amplitude of disturbance;

p is frequency.

After the corresponding transformations, the equations for the torques in the sections are obtained:

$$\begin{gathered}
M_{\kappa p\,1} = C_1\,(\varphi_2-\varphi_1),\\
M_{\kappa p\,2} = C_2\,(\varphi_3-\varphi_2),\\
M_{\kappa p\,n-1} = C_{n-1}\,(\varphi_n-\varphi_{n-1}).
\end{gathered} \quad (4)$$

After substituting the values φ_1, φ_2, ..., φ_n, in formulas (2.4), the following equations are obtained:

$$\begin{gathered}
M_{\kappa p\,1} = C_1\,(a_2-a_1)\,Cos\,pt,\\
M_{\kappa p\,2} = C_2\,(a_3-a_2)\,Cos\,pt,\\
M_{\kappa p\,n-1} = C_{n-1}\,(a_n-a_{n-1})\,Cos\,pt.
\end{gathered} \quad (5)$$

For the frequency analysis of the vibrational transmission system of a wheeled forestry tractor, the transmission and frequency functions use is of interest.

The differential equations system is the torsional vibrations equations. Solving this system of equations with zero right-hand side, the natural oscillations of each inertial mass can be found. The solution of the system with the right-hand side, which describes the disturbing moment applied to the elements, allows determining the forced oscillations of the system under the action of disturbing forces that periodically change over time. These vibrations increase the dynamic transmission loading.

Acknowledgements

The result of solving these equations, taking into account external disturbances, is presented in a tabular form. The table shows the engine natural frequencies and disturbance frequencies of the “engine - transmission - tractor - wood bundle” system at the theoretical speed of the skidding system in various gears for seven mass dynamic design scheme. The dominant operating range of the engine crankshaft rotational speed of a wheeled forestry tractor is from 1800 ... 2000 min^{-1}, in short-term modes - 1200 min^{-1}, and in cases of short-term engine overloads, the speed can be reduced to 800 min^{-1}.

The operator’s desire to obtain maximum productivity results in the fact that the main part of the transport operations machine time, the engine works when the control lever is set to the maximum fuel supply. This is accompanied by a relatively stable rotational speed of the engine crankshaft, regardless of operating conditions and tractor power saturation.

In order to assess the engine speed the crankshaft speed utilization is defined:

$$K_n = n_e / n_{ен}, \tag{6}$$

where n_e is mathematical expectation of engine crankshaft speed;

$n_{ен}$ is engine crankshaft speed at rated power.

The theoretical movement speed is calculated with the engine speed utilization coefficient $K_П = 0,8$.

It is known that when the disturbing force frequency coincides with the natural oscillations frequency of the system, resonant oscillations occur. These oscillations are maintained indefinitely due to the action of a disturbing force and, therefore, are of great practical importance [5,7].

The amplification factor β is the ratio [4]

$$\beta = \frac{a}{a_0} \tag{7}$$

where a is the steady-state amplitude of the disturbing frequency strength,

ω; a_0, $a_a = h/p^2$ is the deviation of the system from the equilibrium position under the constant force action;

$h = H / m$, H is the amplitude;

m is the mass.

Table 1 – Eigen frequencies values of the system "engine-transmission-tractor-bundle» of the 7 mass system

Parameters	Transmissions			
	I	II	III	IV
Eigen frequency mode, rad/s	4,7	4,703	4,703	4,703
	8,05	8,903	7,873	5,571
	58,32	56,4	33,34	28,85
	142,8	243,9	447,77	802,6
	304,1	365,4	530,31	877,7
	4632	4190	3940	3460

For small damping coefficient values, the maximum value of the coefficient β is observed at a point lying very close to the resonance; therefore, it is quite acceptable to take its value at resonance as the maximum value of the coefficient β.

Therefore, the analysis of the “engine – transmission – tractor – bundle” system eigen frequencies and the lugs synchronous excitation frequencies should be carried out not only at the resonance, when $\omega/P = 1$, but also in the resonance area, in which the coefficient β has a significant value; in this research a resonance area (0,75÷1,25) ω/P is taken.

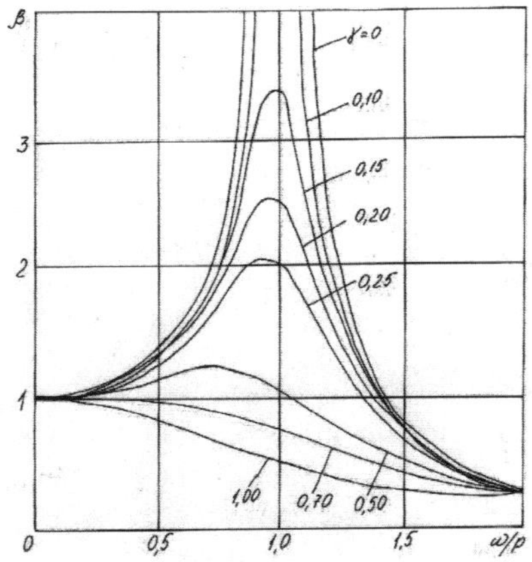

Figure 2. The resonance curve of the oscillatory circuit:
P is eigen frequencies; *ω is* imposed frequency; *β* gain factor

Proin pharetra nonummy pede. Mauris et orci. Aenean nec lorem. In porttitor. Donec laoreet nonummy augue. Suspendisse dui purus, scelerisque at, vulputate vitae, pretium mattis, nunc. Mauris eget neque at sem venenatis eleifend. Ut nonummy.

Conclusions

Analysis of the eigen frequencies and the system exciting frequencies allows to draw the following conclusions:

- the engine-and-transmission installation can operate in the resonance zone in lower gears: the eigen frequency is 56.54 rad/s with a crankshaft speed in the range of 1200 min^{-1} in first gear and 140.6 rad/s with a speed of 2000 min^{-1} in second gear during the interaction of the wheels lugs with the bearing surface;
- the engine-and-transmission installation can operate in the resonance zone in higher gears (3rd and 4th gear) with a crankshaft rotation frequency in the range of 1200 ... 2200 min^{-1}.

In order to exclude resonant modes in the transmission of a wheel skidder, the possibility of installing elements with greater flexibility is considered.

References

[1] Anisimov G.M. Operating conditions and loading of the skidder transmission. Moscow. Forest industry, 1975 p 170.
[2] Anisimov G.M. The operational efficiency of skidders. Moscow. Forest industry, 1990 p 208.
[3] Technical level analysis of tractors and proposals development to improve it. Designs' analysis of modern domestic and foreign forestry tractors: Report on research work / SZF NATI Vyritsa, 1987 p 80.
[4] Study of the perspective wheel tractor TLK-4-01 operating modes and transmission load: Report on research work FTA St. Petersburg., 1996.
[5] Loitsapsky L.G., Lurie A.I. The course of theoretical mechanics. Moscow. Mechanical Engineering, 1985 p 638.
[6] Taubes L.E. On the issue of the oscillations passing through the torque converter. In the book: The main directions of scientific research and experimental design work on the rational parameters and operational qualities of automotive equipment selection. Mn., 1976 p 67-68.
[7] Investigation of the technical solutions and parameters influence on the traction and speed properties of promising wheeled short log tracks. Report on research work LTA 1.12.036; N GR O191OO52777 St. Petersburg, 1991 p 55.

IOP Conf. Series: Materials Science and Engineering **695** (2019) 012023 doi:10.1088/1757-899X/695/1/012023

Composition research of generator gas for application as motor fuel

S.A.Plotnikov[1], A.L.Biryukov[2], A.S.Zubakin[1]

[1]Vyatka State University
[2]Vologda State Dairy Academy n.a. N.V. Vereshchagin

Abstract. The problem of the lack of currently approved methodology for determining chemical composition of generator gas and the effect of gas composition when using it as a motor fuel on the operational and environmental performance of an internal combustion engine is discussed. A method for determining chemical composition of generator gas has been developed and tested, in which it is proposed to measure mass concentrations (volume fractions) of the determined components in the generator gas by the indicator (linear-colour) method. Experimental studies were carried out to determine chemical composition of the generator gas according to the developed technique. The studies were carried out on an experimental gas generator of original design, the novelty of which is confirmed by a patent for the invention. Charcoal and wood chocks were used as raw materials for production of generator gas. The composition of the generator gas in a parametric gas generator depends on the type of fuel (raw material), the temperature in the gasification zone, the number and spatial location of the injectors included in the operation and varies in the content of combustible components: H2 from 7% to 18%; CnHm - from 0.1% to 0.3%; CO - from 18% to 31%; CH4 - from 2% to 2.5%. The content of ballast in the generator gas also changes, so the O2 content is from 0.2% to 1.9%; N2 - from 48% to 57.3%; CO2 - from 3.2% to 10%. Experimental studies have confirmed the sufficient accuracy of the proposed methods.
Key words: technique, generator gas, chemical composition, linear coloristic method, indicator method, experimental studies.

Introduction

Generating gas is one of the most promising types of alternative fuels due to its renewability and attitude to the so-called "local" fuels.

Despite this, at present, generator gas is not a standard fuel and, therefore, there are no approved standard methods for determining its composition. In this regard, developing methods to determine composition of the generator gas and their testing is very important [1-12, 16-18].

It should be noted that there are a certain number of existing methods to determine composition of various gases in the atmosphere, industrial and other emissions, etc. [13,14,15 and others]. However, they all have different tasks and are not standard for determining composition of the generator gas.

One of the problems at the same time is the insufficient knowledge of the chemical composition of the generator gas itself, obtained under various conditions and process parameters, from various raw materials, and especially its mixtures with other fuels, as well as its effect when used as a motor fuel on operational and environmental performance of an internal combustion engine [1-6,8,9]

Chemical and element-wise composition of the generator gas is an important characteristic and determines the rate of combustion of the air-fuel mixture, temperature and maximum pressure in the cylinder of the internal combustion engine, and therefore various power and environmental indicators. So, for example, with an increase in the content of hydrogen or heavy hydrocarbons in generator gas, a sharp increase in maximum pressure in the cylinder occurs. When comparing the chemical composition of gasoline and generator gas, it can be noted that an increase in hydrogen leads to an increase in detonation resistance. The presence of a large amount of nitrogen in the composition of the generator gas should reduce the rate of combustion of the mixture. [1-6,8,9].

The composition of the generator gas differs significantly from the composition of traditional hydrocarbon fuels, which causes changes in the operational and environmental performance of the engine converted to generator gas and justifies the need to compare their chemical compositions [1-6].
Also, taking into consideration, the need to increase the reaction rate of an internal combustion engine, operating on a generator gas to an increasing load, the use of generator gas in a mixture with other light fuels in appropriate modes becomes relevant with gasoline, the combustion processes of which also require study [3,5-7].

Description of the ways to determine composition of the generator gas
It is suggested to use a method for determining chemical composition of generator gas, with the development of which some requirements were taken into account, which, in our opinion,were close to the standard STO MI 2606-2018 “Method for measuring mass concentrations (volume fractions) of carbon monoxide, nitrogen oxides, sulfur dioxide, hydrogen sulfide, formaldehyde, acrolein in the air of the working area, industrial emissions by the indicator (linear-coloristic) method using chemical gas detectors of the GC-E type ”[13].

It is proposed to carry out the measurement of mass concentrations (volume fractions) of the determined components in the generator gas by the indicator (linear-coloristic) method. The indicator measurement method is based on measuring the length of the reagent layer, placed in transparent indicator tubes, which has reacted with the detected component in the test sample and has changed color. The length of the reacted filler layer depends on the content of the component being determined and the volume of the sample taken for analysis. The content of the determined component in the analyzed sample can be determined by the scale applied to the indicator tube. When measuring the mass concentration of NO2, NOX and CO, the atmospheric pressure, the temperature of the generator gas at the inlet to the indicator tube and the overpressure or vacuum at the sampling point are measured.

Measurements are proposed to be carried out in the following order. The indicator tube 7 filled with reagent is installed in the connector of the aspirator 8 (Fig. 1), and the opposite end of the indicator tube is connected to the duct 1 with the test generator gas through a gas intake tube with a condensate collector 2 and a tee 4 with a thermometer.

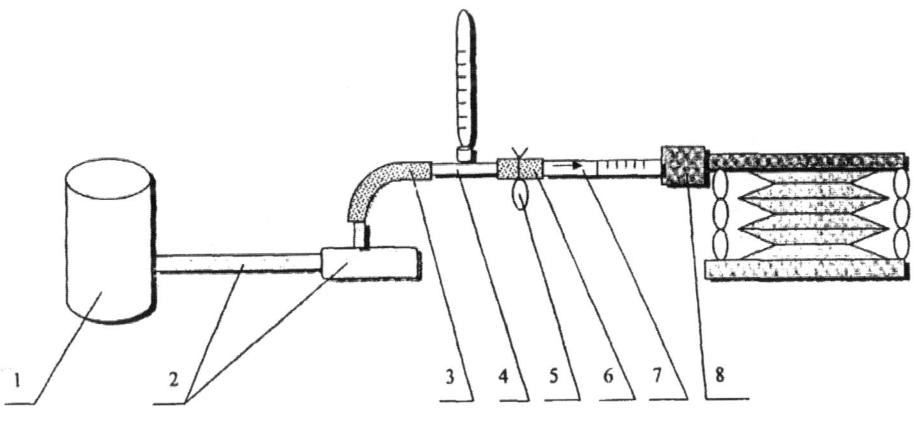

Figure 1. Installation for determining composition of generator gas: 1 - gas duct; 2 - gas intake pipe with condensate collector; 3 - rubber tube; 4 - thermometer tee; 5 - clamp; 6 - rubber tube; 7 - indicator tube; 8 - aspirator [13]

A general view of the instruments used to determine the composition of the generator gas is shown in Figures 2 and 3.

IOP Conf. Series: Materials Science and Engineering **695** (2019) 012023 doi:10.1088/1757-899X/695/1/012023

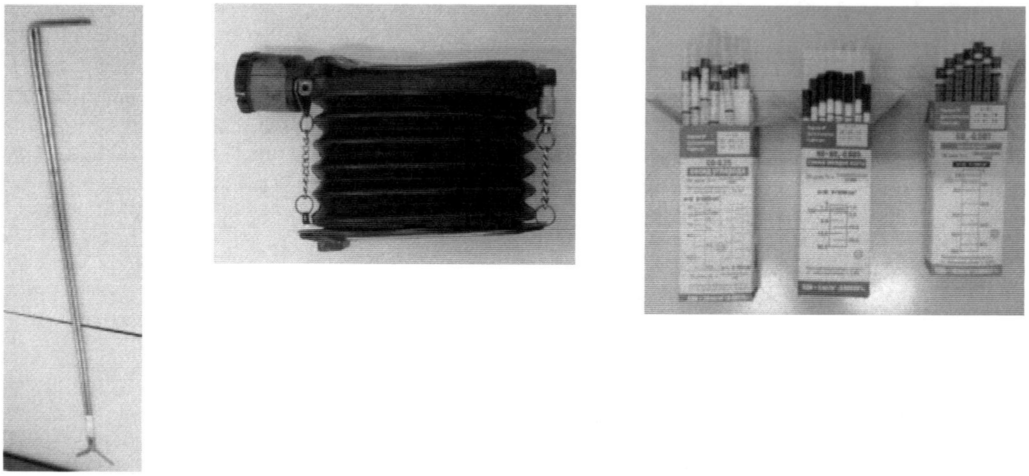

a) *b)* *c)*

Figure 2. Measuring instruments: intake pipe (a), aspirator (b), indicator tubes (c)

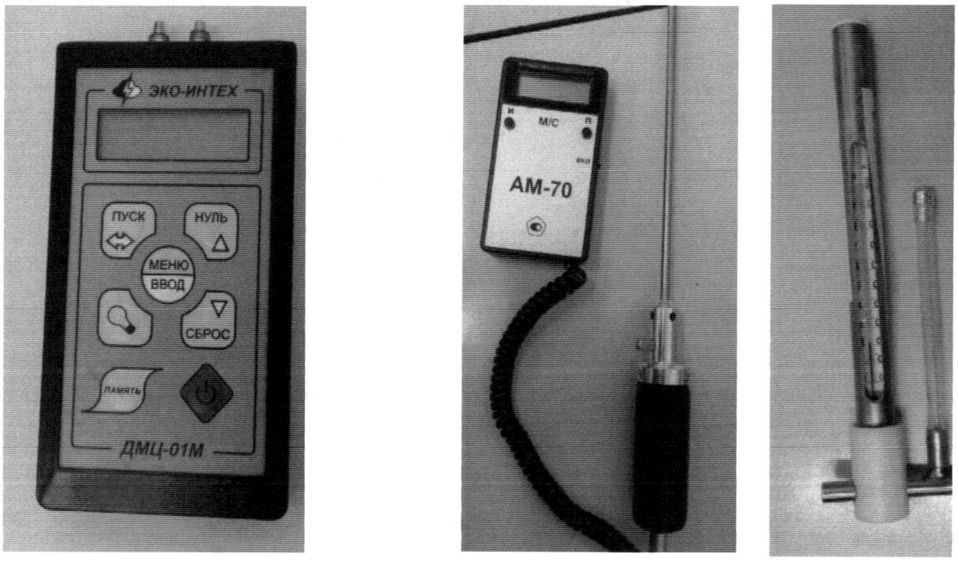

a) *b)* *c)*

Figure 3. Measuring instruments: differential pressure gauge DMC-01M (a); AM-70 anemometer (b); thermometer TM-100 (c)

Before conducting research on the composition of the generator gas, the gas intake path is washed with ten times the volume of the investigated generator gas by using an aspirator when connecting it to the tee of the thermometer with the indicator tube removed. Before disconnecting the aspirator, the gas intake must be sealed with a clamp.

Sampling of the generator gas is carried out at a steady state operation of the gas generator and internal combustion engine. It is necessary to use a new indicator tube, in which it is necessary to break off both ends without violating the provisions of the filter pad and reagent layer, for example, using a knife on the aspirator body. Then the indicator tube is installed with an unmarked end into the aspirator connector, and the end of the marked tube is connected to the tee 4. The direction of movement of the generator gas sample should coincide with the direction of the arrow marked along the scale of the indicator tube.

To determine the content of the analyzed component, the aspirator bellows is quickly compressed with one or two hands to the stop, and then released. The aspirator should open until the restriction chains are fully tensioned. In this case, there should be a sharp movement of the lever associated with the valve. Then, the indicator tube is disconnected from the aspirator and 1-2 minutes later the mass concentration (volume fraction) of the analyzed component of the generator gas is determined according to the calibration scale applied to the indicator tube. In this case, it is necessary to take the value that corresponds to the boundary of the layer that has changed colour. In the case where an uneven boundary of the coloured layer is observed, an average value should be taken.

In the process of determining mass concentrations (volume fractions) of the analyzed components, it is necessary to fix the values of the temperature of the generator gas, as well as atmospheric pressure and humidity.

In preparation for making measurements in the laboratory according to GOST 15150-69, the following conditions must be fulfilled: the ambient temperature must be within 20 ± 5 ° C; relative air humidity - not more than 80% at a temperature of 25 ° C; atmospheric pressure - 84.0 - 106.7 kPa (630 - 800 mm Hg). In the process of determining the mass concentrations of the analyzed components, it is necessary to comply with the conditions adopted according to [14] and indicated in table 1.

Table 1 - Requirements for the conditions to measure chemical composition [14]

Parameter	***Value***
Environment temperature , ° C	0 … 50
Atmospheric pressure, kPa	80 … 113,3
Humidity at temperature of 25 ° C, %	up to 100
Generator gas temperature at the sampling point, ° C	up to 250
Overpressure (depression) of the analyzed gas sample, kPa	not more than± 5
Relative humidity at a temperature of 100 ° C, %	up to 10
Sample temperature of the generator gas at the inlet to the indicator tube, ° C	up to 50

Calculation of volume fractions of the component content in the generator gas should be taken according to the following formula:

$Cv = A / Ka$

where Cv is the volume fraction of the analyzed component,%;

A is the value determined on the scale of the indicator tube;

Ka is a coefficient taken depending on the number of strokes of the aspirator and the volume of generator gas passing through the indicator tube. It is recommended to take into account the data [13].

Recalculation of volume fractions of the analyzed component in mass concentrations can be carried out according to the formula:

$Cm = \kappa \times Cv$

where Cm is the mass concentration of the analyzed component, mg / m3;

k is the conversion factor.

The conversion factor should be taken depending on the analyzed component according to [13].

Experimental research

To test the proposed methodology and in full accordance with it, experimental studies of the composition of the generator gas have been carried out. The studies were performed on generator gas obtained by using charcoal ($\acute{\omega}$ = 20%) and wood chocks ($\acute{\omega}$ = 20%) as raw materials on an experimental parametric gas generator of the original design (Fig. 4) [19].

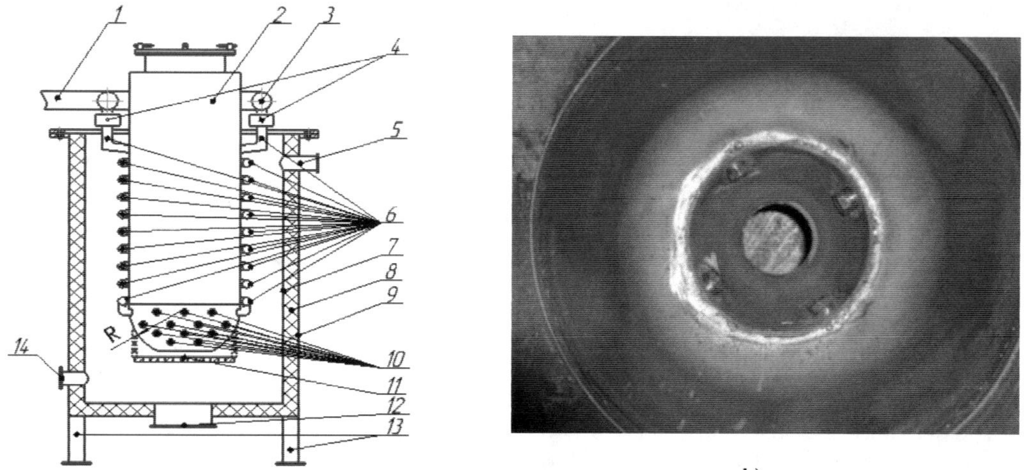

a) b)

Figure 4. Parametric gas generator with regulation of the working process in the volume of the gasification chamber: a) - diagram; b) - view of the tuyere belt:
1 -air pipe; 2 -gasification chamber; 3 - air collector; 4 - electromagnetic valves; 5 - gas outlet; 6 - tubes; 7 - gas tank; 8 - thermal case; 9 - protective casing; 10 - tuyere belt; 11 - ash pan grate; 12 - ash pan hatch; 13 - supports; 14 - technological hatch.

This gas generator makes it possible to obtain generator gas of various chemical composition during various gasification processes (direct, reverse and horizontal), which is due to the peculiarity of gasification chamber 2 and the presence of controlled tuyeres 10.

The results of experimental studies are presented in table 2.

Table 2 – Chemical composition of the generator gas

Names of raw materials and gasification process parameters	**Combustible components,%**				**Ballast,%**		
	H_2	C_nH_m	*CO*	CH_4	O_2	N_2	CO_2
Wood chocks, ώ = 20%	14-18	0,1-0,3	18-22	2,1-2,5	1,4-1,8	48-52	8-10
Charcoal, transverse gasification process	7,0-8,0	0,2-0,3	30,3-31,0	2,0-2,5	0,2-0,3	55,0-57,3	3,2-3,5
Wood chocks, reversed process	16,1-16,8	0,2-0,28	20,1-21,2	2,1-2,4	1,6-1,9	48,7-50,4	8,2-9,3

As is seen from the table, the composition of the generator gas in the parametric gas generator varies widely enough and depends on the type of fuel, the temperature in the gasification zone, the number and spatial location of the injectors included in the operation.

Analyzing the chemical composition of the generator gas taken using an experimental gas generator and comparing the obtained values of the chemical composition of the generator gas with the same values given in the technical literature, we can conclude that the proposed method allows us to determine the chemical composition of the generator gas with an allowable error.

Conclusions

1. At present, there are no approved standard methods for determining composition of generator gas.
2. Composition of the generator gas, obtained under various conditions and process parameters, from various raw materials, and especially its mixtures with other fuels, which is very important due to the need to increase the reaction rate of the internal combustion engine to an increasing load, as well as the influence of the composition of the generator gas when using it as a motor fuel, the operational and environmental performance of an internal combustion engine are not sufficiently studied.

3. Composition of the generator gas in a parametric gas generator depends on the type of fuel (raw material), the temperature in the gasification zone, the number and spatial location of the injectors included in the operation and varies in the content of combustible components: H2 from 7% to 18%; CnHm - from 0.1% to 0.3%; CO - from 18% to 31%; CH4 - from 2% to 2.5%. The content of ballast in the generator gas also changes, so the O2 content is from 0.2% to 1.9%; N2 - from 48% to 57.3%; CO2 - from 3.2% to 10%.

4. Proposed methodology for determining the chemical composition of the generator gas based on the indicator method allows to determine the chemical composition of the generator gas with an allowable error .

References

[1] Analysis of the combustion process of the generator gas and its mixture with gasoline in a piston engine. / Plotnikov S.A.,. Zubakin A.S. // Dvigatelestroyenie. - 2018. - No. 3. - p 14–18.
[2] On some features of the engine gas operation / Plotnikov S.A., Kantor P.Ya., Zubakin A.S., Vtyurina M.N. // Works of NSTU n.f. R.E. Alekseev. 2019.No 2 (125). p. 192-196.
[3] Improving the smoothness and accuracy of regulating the performance of the gas generator for the internal combustion engine of mobile power tools and stationary power plants / Biryukov A.L., Kartoshkin A.P., Shushkov R.A., Zefirov I.V. // Proceedings of the International Academy of Agricultural Education - 2018. - No. 41 - Volume 2 – p..91-94.
[4] Improving energy and environmental performance of ICE when working on generator gas / Plotnikov SA, Zubakin AS In the collection: Society. Science. Innovation (NPK-2017) collection of articles. All-Russian annual scientific-practical conference. Vyatka State University. 2017.p. 1828-1834.
[5] Gas generator: Patent No. 168538 Russian Federation, IPC C10J3 / 20 (2006.01) / Biryukov A.L., Plotnikov S.A., Shushkov R.A., Zubakin A.S .; Applicant and patent holder FGBOU VO Vologda GMHA. No. 2016113671; Dec. 08.04.2016; publ. 02/08/2017 Bull. Number 4
[6] Development of gas generator for use on mobile energy / Biryukov A. L., Plotnikov S. A., Kartoshkin A. P., Zubakin A. S., Shushkov R. A. // Proceedings of the International Academy of Agricultural Education - 2017. - No. 35 - p.15-18.
[7] Dasappa, S., and Sridhar, H. V. 2013. Performance of diesel engine in a dual fuel mode using procedure gas for electricity power generation. Int. J. Sustain Energy 32:153–168.
[8] Uma, R., C. T. Kandpal, and N. V. V. Kishore. 2004. "Emission Characteristics Electricity Generation System in Diesel Alone and Dual Fuel Modes." Biomass and Bioenergy 27: 195–203.
[9] Guskov V.F., Sukhov S.P., Kozlov S.A. Results of the cycle study for the engine running on generator gas // Successes of modern science, 2012. - No. 6. - p. 27-28.
[10] Haggith D, Sobiesiak A, Miller L, Przybyla G. Experimental indicated performance of a HCCI engine fueled by simulated biomass gas. SAE International, Technical Paper. 2010.
[11] Nwafor, O. M. I. 2000a. "Effect of Choice of Pilot Fuel on the Performance of Natural Gas in Diesel Engine." Renewable Energy 21: 495–504.
[12] G Sridhar, H V Sridhar, S Dasappa, P J Paul, N K S Rajan and H S Mukunda. Development of Producer Gas Engines. //Combustion Gasification and Propulsion Laboratory. - Department of Aer-ospace Engineering. - Indian Institute of Science. - Bangalore, India.
[13] STO MI 2606-2018. Methodology for measuring mass concentrations (volume fractions) of carbon monoxide, nitrogen oxides, sulfur dioxide, hydrogen sulfide, formaldehyde, acrolein in the air of the working area, industrial emissions by the indicator (linear-coloristic) method using chemical gas detectors of the GH-E type "/ Ekaterinburg: "Prombezopasnost" , 2018.- 23 p..
[14] MVI-1-06. Methodology for measuring the content of nitrogen oxides, carbon monoxide and oxygen using a set of indicator tubes in the organized emissions of boiler houses, TPPs and state district power stations operating on natural gas "/ St. Petersburg, 2006. - 14 p..
[15] MVI-2-05. Methodology for measuring the mass concentration of vapors of oil products, technical mixtures and solvents in sources of atmospheric pollution and in the air of the working zone using indicator tubes / St. Petersburg, 2006, 17 p.
[16] Gnanam G, Haggith D, Sobiesiak A. A novel in cylinder fuel reformation approach to control HCCI engine combustion on-set. PTNSS-2009-SS3-C145 Combust Engines 2009;3:37-48.
[17] Shudo T, Hamada H. Hyrdrogen as an ignition-controlling agent for HCCI combustion engine by suppressing the low-temperature oxidation. Int J Hydrogen Energy 2007;32:3066-72.

[18] Achilles M., Ulfvik J., Tuner A.M., Johansson B., Ahrenfeldt J., Henriksen U.,Schauer F.X., HCCI gas engine: evaluation of engine performance, efficiencyand emissions ecomparing producer gas and natural gas, SAE Technicalpaper 2011-01-11196.
[19] Gas generator: Patent No. 2575536 Russian Federation, IPC C10J3 / 20 (2006.01) / Ostretsov V.N., Zubakin A.S., Palitsyn A.V., Korotkov A.N., Kipriyanov F.A., Rassvetalov A .FROM.; patent holder A. Palitsyn No. 20144132727/05; declared 08/07/2014; publ. 02/20/2016 Bull. Number 5

IOP Conf. Series: Materials Science and Engineering **695** (2019) 012024 doi:10.1088/1757-899X/695/1/012024

Neural networks modification for solving the traffic signs detection problem

Devyatkin A.V., Filatov D. M.
Department of ACS, SPbSETU "LETI", St. Petersburg, Russia
E-mail: avdevyatkin@etu.ru, dmfilatov@etu.ru

Received xxxxxx
Accepted for publication xxxxxx
Published xxxxxx

Abstract. The paper deals with the implementation of the traffic signs detection model on the deep neural network basis. The complexity of the problem lies in the detected objects size. The paper considers the use of modern approaches to the deep neural networks Mobile Net v2 and Tiny YOLO v3 implementation to solve the detection problem in real time. Also, a modification of the considered networks is proposed, which allows increasing the Average Precision index by more than 20%. For the networks training GPGPU NVIDIA 1080 and a signs images set called Russian traffic sign images dataset - RTSD which includes more than 15,000 frames for training and more than 3,000 frames for carrying out testing were used.
Keywords: deep neural network, detection, YOLO, MobileNet

1. Introduction

Modern working methods with images and developments in the field of algorithms based on training allow determining various objects with a sufficiently high accuracy and highlighting their location in the frame. In the initial stages, classical methods were used, such as the Haar classifier, the template method, the Viola-Jones method. Today, due to the development of special hardware platforms, modern neural network architectures allow both increasing the accuracy and detection methods speed. Such architectures are SSD [1], YOLO [2], Fast R-CNN [3], Mask R-CNN [4], FPN [5]. Each of the approaches has its own specifics, but the general trend in the recognition problem is testing on generalized data sets specially prepared for evaluation and comparison of detection algorithms. The most common datasets are COCO [6], Pascal VOC [7].

In the field of computer vision for the unmanned vehicles creation, the researchers highlight one of the important tasks - traffic signs detection, which allows the control system when using unmanned mode or driver assistance mode to identify signs as objects with information and take this information into account when planning or controlling the traffic trajectory. Despite the fact that in the future, with full road traffic automation, the need for the road signs use may completely disappear in the coming years, when on public roads vehicles with both autonomous control and under human control will be used, the problem of road signs recognition will not lose its relevance.

To solve the detecting signs problem with the use of methods based on training, special data sets are used, which contain marked data corresponding to the task specifics. In the development of the road signs detector such dataset as German set of signs [8], Belgian set of signs [9], Swedish set of signs [10] and others can be used.

The similarity of general and specialized datasets with road signs is that the annotation is made with the area indication in the image. However, the road signs detecting task is complicated by the fact that the size of the sign on the frame is often relatively small compared to the General dataset, where objects can occupy half a frame. The sign image can occupy several tens of pixels in width per Full HD image, which determines the specifics of this task.

2. Neural networks for detection

Recognition approaches based on deep neural networks can be evaluated by a number of indicators. In this paper two of the most interesting indicators are identified:

-Operation speed,

- Detection accuracy.

Previously, the development trends were aimed at increasing the detection methods accuracy, but today the hardware of the calculations allows getting the work of methods in real time, which is especially important when driving unmanned vehicles. The following two architectures stand out as the solutions: SSD and YOLO. Both architectures are one-pass detection methods, for which the input is the image, and the result is predictions on the object localization with the indication of the class in the image.

This paper uses the Yolo 3 architecture [2] as one of the real-time detectors with the classic Tiny configuration (a smaller version of the original network). Also for comparison and evaluation the basis of Tiny YOLO is replaced by MobileNet2 architecture [11], which consists of special blocks aimed at reducing the number of parameters for network operation, which has a positive impact on learning and memory consumption.

YOLO architecture can be divided into two blocks:

- the basic convolution unit,
- the output prediction layer(-s).

The main convolution unit is the main component for the selection and processing of features in the image. The number of filters in the prediction output layer and the size of the layer are set according to the number of predicted feature classes and the number of cells into which the image is split.

According to this view, in the classic Tiny YOLO architecture, it is possible to replace the main feature allocation unit with the MobileNet2 architecture network. A summary view of the architectures is presented in tables 1 and 2. Layer designations are deciphered as follows:

- Conv-Convolution,
- Concat-Concatenation,
- CBR-Convolution + Batch Normalization + Relu,
- MP-Max Pooling,
- US - Up sampling,
- IRB - Inverted residual block [11]

Table 1. Tiny YOLO network description

#	Input#	Layer, output size	Number of filters, the size of the convolution kernel
0		Input, 416x416	
1	0	CBR + MP, 208x208	16, 3
2	1	CBR + MP, 104x104	32, 3
3	2	CBR + MP, 52x52	64, 3
4	3	CBR + MP, 26x26	128, 3
5	4	CBR + MP, 13x13	256, 3
6	5	CBR, 13x13	512, 3
7	6	CBR, 13x13	1024, 3
8	7	CBR, 13x13	256, 1
9	8	CBR, 13x13	512, 3
10	9	Conv, 13x13	18, 1
11	8	CBR + US, 26x26	128, 1
12	11, 4	Concat + CBL, 26x26	256, 3
13	12	Conv, 26x26	18, 1

Table 2. Mobile Net 2 description

#	Input#	Layer, output size	Number of filters, the size of the convolution kernel
0		Input, 416x416	
1	0	CBR + MP, 208x208	32, 3
2	1	IRB, 208x208	16, 3
3	2	IRB x2, 104x104	24, 3
4	3	IRB x3, 52x52	32, 3
5	4	IRB x4, 26x26	64, 3
6	5	IRB x3, 26x26	96, 3
7	6	IRB x3, 13x13	160, 3
8	7	IRB, 13x13	320, 3
9	8	CBR, 13x13	1280, 1
10	9	Conv, 13x13	18, 1

In this paper, training based on the Russian roads signs set (RTSD) is used [12]. This dataset contains more than 15,000 frames from the road for training and more than 3,000 frames for testing algorithms. For training, the Adam method was used with a learning factor of 0.001. Data expansion was performed using augmentation by the following methods:

- random shift,
- random turn,
- random horizontal reflection,
- normally distributed noise,
- color distortions in HSV space.

For evaluation purposes, the Average Precision (AP) metrics are used with a threshold value of the Jakarta (or Intersection Over Union) equal to 0.5 and the speed of work in frames per second (FPS) of the network on the test data set. The input image size for both architectures is 416x416 pixels. Both network configurations use three reference rectangles per output layer. The Tiny YOLO network uses output layers of size 13 and 26, while MobileNet 2 uses a standard output layer of size 13. The number of output layers is set according to the formula:

$$F = N * (4 + 1 + C),$$

where F is the number of prediction layer filters, N is the number of reference rectangles per output layer, and C is the number of predicted classes. In this case: C = 1, N = 3, F = 18.

The performance indicators of trained networks are presented in table 3.

Table 3. Network accuracy indicators

Architecture type	AP, %	FPS
Tiny YOLO	30,8	96,9
MobileNet2	29,6	101.3

As can be seen from the indicators presented in table 3, the proposed standard network configurations, focused on a common set of medium and large objects, cope with the detection task with low accuracy.

3. Network modification and results

Based on the idea that the minimum cell size of the output grid in the Tiny YOLO network is 16 pixels, it can be assumed that adding outputs from the 8 pixel grid sizes should increase the resolution of the model and increase the accuracy of detecting small objects, as this will allow the model to analyze the input image in more detail.

IOP Conf. Series: Materials Science and Engineering **695** (2019) 012024 doi:10.1088/1757-899X/695/1/012024

To improve the performance, the following modifications of the source architectures are proposed to improve detection accuracy (tables 4 and 5).

Table 4. Tiny YOLO modification for small objects

#	Input #	Layer, output size	Number of filters, the convolution kernel size
0		Input, 416x416	
1	0	CBR + MP, 208x208	16, 3
2	1	CBR + MP, 104x104	32, 3
3	2	CBR + MP, 52x52	64, 3
4	3	CBR + MP, 26x26	128, 3
5	4	CBR + MP, 13x13	256, 3
6	5	CBR, 13x13	512, 3
7	6	CBR, 13x13	1024, 3
8	7	CBR, 13x13	256, 1
9	8	CBR, 13x13	512, 3
10	9	Conv, 13x13	18, 1
11	8	CBR + US, 26x26	128, 1
12	11, 4	Concat + CBL, 26x26	256, 3
13	12	Conv, 26x26	18, 1
14	11	CBR + US, 52x52	128, 1
15	14, 3	Concat + CBL, 52x52	128, 3
16	15	Conv, 52x52	18, 1

Table 5. MobileNet 2 modification for small objects

#	Input#	Layer, output size	Number of filters, the convolution kernel size
0		Input, 416x416	
1	0	CBR + MP, 208x208	32, 3
2	1	IRB, 208x208	16, 3
3	2	IRB x2, 104x104	24, 3
4	3	IRB x3, 52x52	32, 3
5	4	IRB x4, 26x26	64, 3
6	5	IRB x3, 26x26	96, 3
7	6	CBR, 26x26	1280, 1
8	7	Conv, 26x26	18, 1
9	6	US, 52x52	
10	9, 4	Concat + IRB x2, 52x52	32, 3
11	10	CBR, 52x52	640, 1
12	11	Conv, 52x52	18, 1

As can be seen, in the presented modifications Tiny YOLO has output layers with the size of 13, 26 and 52 cells, and MobileNet2 - only 26 and 52 cells, omitting the possibility of analyzing the image using a grid with 13 cells. This approach assumes that the modified MobileNet2 architecture will weakly detect large objects that are not present in the data set as such.

The learning outcomes and assessments in the test sample are presented in Table 6.

Table 6. Accuracy indicators of the modified networks

Architecture types	AP, %	FPS
Tiny YOLO	48,7	48,2
MobileNet2	66,0	45,9

As can be seen from the indicators, the modification of the Tiny YOLO network allowed to greatly increase the accuracy of the network operation with a corresponding decrease in the operation speed. The presence of an output layer with a size of 13x13, focused on the analysis of sufficiently large objects, presumably reduces the model efficiency, which is confirmed by greater accuracy in the modification MobileNet2, which has no large objects analysis.

Decrease in the speed of both modifications does not exceed the limits of real time operation and is relatively higher than the speed of most modern cameras, for this reason they can be considered acceptable.

4. Conclusion

This paper presents an approach to modify the deep neural network architecture in order to increase the detection accuracy when working with small objects. The presented results confirm the hypothesis that increasing the resolution of the output layer grid (the output layer size) increases the ability of the network to analyze small objects, which is particularly important in working with remote traffic signs on the road. The considered approach allows to apply General purpose neural networks and to increase detection accuracy at work with certain ranges of detectable objects.

Further work involves the neural networks study in terms of the number of filters parameters and sizes in order to compensate for the loss in speed without accuracy loss in detecting traffic signs.

References

[1] Liu, Wei et al. “SSD: Single Shot MultiBox Detector.” Lecture Notes in Computer Science (2016): 21–37. Crossref. Web.

[2] J. Redmon and A. Farhadi. Yolov3: An incremental improvement. arXiv preprint arXiv:1804.02767, 2018.

[3] Ross Girshick. Fast R-CNN. arXiv preprint arXiv:1504.08083, 2015.

[4] Kaiming He and Georgia Gkioxari and Piotr Dollár and Ross Girshick. Mask R-CNN. arXiv preprint arXiv:1703.06870, 2017.

[5] Tsung-Yi Lin and Piotr Dollár and Ross Girshick and Kaiming He and Bharath Hariharan and Serge Belongie. Feature Pyramid Networks for Object Detection. arXiv preprint arXiv:1612.03144, 2016.

[6] Lin, Tsung-Yi et al. “Microsoft COCO: Common Objects in Context.” Lecture Notes in Computer Science (2014): 740–755. Crossref. Web.

[7] Mark Everingham, Luc Gool, Christopher K. Williams, John Winn, and Andrew Zisserman. 2010. The Pascal Visual Object Classes (VOC) Challenge. Int. J. Comput. Vision 88, 2 (June 2010), 303-338. DOI=http://dx.doi.org/10.1007/s11263-009-0275-4

[8] J. Stallkamp, M. Schlipsing, J. Salmen, C. Igel, Man vs. computer: Benchmarking machine learning algorithms for traffic sign recognition, Neural Networks, Available online 20 February 2012, ISSN 0893-6080, 10.1016/j.neunet.2012.02.016. (http://www.sciencedirect.com/science/article/pii/S0893608012000457)

[9] Radu Timofte*, Markus Mathias*, Rodrigo Benenson, and Luc Van Gool, Traffic Sign Recognition - How far are we from the solution?, International Joint Conference on Neural Networks (IJCNN 2013), August 2013, Dallas, USA.

[10] Fredrik Larsson and Michael Felsberg , Using Fourier Descriptors and Spatial Models for Traffic Sign Recognition , In Proceedings of the 17th Scandinavian Conference on Image Analysis, SCIA 2011, LNCS 6688, pp. 238-249. bib , doi:10.1007/978-3-642-21227-7_23

[11] Mark Sandler and Andrew Howard and Menglong Zhu and Andrey Zhmoginov and Liang-Chieh Chen. MobileNetV2: Inverted Residuals and Linear Bottlenecks. arXiv: 1801.04381. 2018.
[12] Shakhuro, V.I. & Konushin, Anton. (2016). Russian traffic sign images dataset. Computer Optics. 40. 294-300. 10.18287/2412-6179-2016-40-2-294-300.

Choice of metal materials for responsible parts and vehicle assemblies operating under the conditions of low temperatures

Y.G. Kabaldin, A.A. Khlybov, M.S. Anosov, D.A. Shatagin, A.A. Golovin, M.D. Scriabin

Nizhny Novgorod State Technical University n.a. R.E. Alekseev

E-mail: ansv-maksim@rambler.ru

Abstract. The purpose of the present study is to evaluate the cold resistance and develop an algorithm for selecting metals that are widely used for manufacturing critical parts and components of vehicles operating in a wide range of low temperatures.
To achieve this goal, the tests were made on impact bending of the studied metals in the temperature range of climatic cold, followed by the study of fractures of metal samples.
The studies were carried out using methods of optical and electron microscopy and fracto-graphic studies. During the study, data were obtained on the change in toughness, the mechanisms of metal destruction for a wide range of low temperatures, as well as the temperature of the viscous-brittle transition of the studied metals.
It has been found that materials with an FCC and HCP lattice exhibit greater resistance to fracture at low temperatures. It is shown that a change in the fracture mechanism is closely related to their crystal structure features and thermal conductivity.
As a result of the research, an algorithm for the selection of cold-resistant materials was developed, and on its basis- a program.

Keywords: impact strength, cold resistance, thermal conductivity, crystalline structure, viscous-brittle transition, vehicles.

Introduction

Modern vehicles operating at low temperatures have a large number of built-in diagnostic systems. Therefore, the application of methods and means of diagnosing a vehicle makes it possible to assess the technical condition of its main components and help its operation be more efficient and effective. Reliability of vehicles operating in the Far North and the Arctic is achieved in another promising way, in particular, due to the correct selection and use of cold-resistant metal materials. Among some important properties of materials operating at low temperatures one can mention toughness and fatigue characteristics that determine the residual life of critical components of vehicles and auxiliary equipment.

In this regard, increasing the reliability and durability of machines in the conditions of operation of the Far North and the Arctic, makes it necessary to develop metal materials with high performance properties, depending on their structural state with stability under both statistical and fatigue loads. Currently, a large number of cold-resistant materials have been developed. Despite their significant diversity, only a small part of them has got the whole range of properties necessary for reliable functioning of technical systems at low temperatures. Depending on operating conditions, metals are subject to brittle or fatigue destruction. At low temperatures, fracture occurs, as a rule in a brittle way. Brittle fracture is more dangerous than fatigue one.

Cold resistant materials are those that retain sufficient viscosity at low temperatures from 0 to - 269 ° C (273 - 4 K). With decreasing temperature, this ability decreases for most metals and alloys. At critical temperatures, shear resistance increases sharply, the metal goes into a brittle state and collapses without any signs of plastic deformation. Resistance to such destruction is called brittle strength, and the property of metals to break brittle with decreasing temperature is called cold brittleness. The inverse concept of cold brittleness is cold resistance [1].

IOP Conf. Series: Materials Science and Engineering **695** (2019) 012025 doi:10.1088/1757-899X/695/1/012025

With brittle fracture at low temperatures, the crack that appears is unstable and grows spontaneously if its length (at a given stress) exceeds a certain critical value and the crack top remains sharp, comparable (along the radius at the peak) with atomic dimensions. In this case, the stresses at the edge of the cracks are sufficient to break the inter-atomic bond. The propagating crack will be bordered by a narrow zone of plastic deformation, the creation of which requires some additional energy [2]. Viscous and brittle fractures are distinguished, which differ in the size of the plastic zone at the crack top [3]. With brittle fracture, the size of the plastic zone at the mouth of the crack is small. With viscous fracture, the size of the plastic zone leading in front of the propagating crack is large, and the crack itself becomes dull at its top. Viscous failure is due to the low propagation speed of the crack. The propagation rate of a brittle crack is very high. Therefore, often brittle destruction is called "sudden", or "catastrophic" destruction.

To assess the cold brittleness of steel, the fracto-graphic control method is also used, based on measuring the fraction (in %) of the fibrous and crystalline structures of impact samples [4].

As a criterion for assessing brittleness, a percentage ratio of the areas of fibrous and crystalline fracture areas is taken. However, in practice, the temperature Tcr., at which the proportion of viscous fracture is 50%, is taken as the criterion of metal viscosity, the less Tcr., the higher the reliability of steel at low temperatures. Tcr was called - the temperature of a viscous-brittle transition.

Lowering the operating temperature is accompanied by an increase in static and cyclic strength [5], with a decrease in ductility and viscosity, and therefore an increase in tendency to brittle fracture. The most important requirement determining the suitability of the material for low-temperature operation of the vehicle is the absence of cold brittleness of metallic materials. Cold brittleness is most characteristic of iron, steel, metals and alloys with BCC lattices.

During operation of the vehicle, its parts in the nodes are subjected to plastic deformation. As a result, elastic deformation energy and heat are stored in the surface layers of the parts.

The energy balance during elastic-plastic deformation of the surface layers of a part in a node can be written as:

$$\Delta E_{м}=\Delta E_{d}+\Delta E_{дис}+\Delta E_{т}, \quad (1)$$

where $\Delta E_{м}$ - level of mechanical energy in the node;

ΔE_{d} elastic strain energy stored in the surface layers of the part;

$\Delta E_{дис}$ - dissipated energy;

$\Delta E_{т}$ - thermal energy, i.e. heat generated in the nodes from friction of parts.

The stored energy ΔE_{d} is determined by the dislocation density. At low temperatures, the motion and propagation of dislocations in metals is limited due to their fixation by impurities. Therefore, the thermo-physical properties of metals are of particular importance in the processes of elasto-plastic deformation of parts. In particular- thermal diffusivity and thermal conductivity. In other words, the ability of a metal to remove heat from the deformation zone at low temperatures is the determining condition for the cold resistance of metals [6]. The fact is that, as was mentioned above (Eq. 1), the energy storage of elastic deformation in the surface layers of a part at low thermal conductivity of the metal and its dissipation will be accompanied by the formation of micro-cracks. Therefore, at low values of heat capacity and thermal conductivity of metals, the occurrence of micro-cracks is the mechanism of dissipation of the stored energy of elastic deformation.

However, the heat capacity of all materials becomes extremely small at cryogenic temperatures (below 200° K), and even a small amount of heat can significantly change the body temperature. The influence of conduction electrons on the specific heat is noticeable only at temperatures close to absolute zero. The heat capacity of solids [7], which changes almost linearly at the beginning of cooling, in the cryogenic region is proportional to the third degree of absolute temperature (Debye law).

In the general case, the thermal conductivity of metals depends both on the electronic structure and on the crystalline structure. Heat is dissipated both by phonons, due to vibrations of the nodes of the crystal lattice, and the motion of free electrons.

The thermal conductivity of relatively pure metals depends mainly on the electronic contribution and, upon cooling, first increases, reaching a maximum, and then quickly drops to zero [7]. In metals with a BCC lattice, the electronic component of thermal conductivity is substantially determined by the presence of impurities in them and the presence of a covalent bond as a result of this, which leads to their low resistance to deformation and brittle fracture at low temperatures.

In this regard, metals with FCC and HCP lattices should exhibit high resistance to brittle fracture. Of structural steels and alloys with an FCC lattice, aluminium alloys (AMg3, AM6, D16), stainless steels, and stainless steels with titanium alloys are widely used. The indicated materials have got higher characteristics of impact strength (temperature of viscous-brittle transition) and fatigue characteristics than carbon steel with a bcc lattice. High characteristics of impact strength and fatigue strength of metals with FCC and HCP lattices are associated with their structure and thermo-physical properties, in particular, thermal conductivity and heat capacity [7].

When choosing a material, first of all, it is necessary to make a comprehensive review of its operating conditions and rank factors affecting the material, according to degree of their influence on the reliability of vehicles and auxiliary equipment, as well as individual units.

Among the important properties of materials operating at low temperatures one can mention their toughness and fatigue characteristics, which determine the structural stability of the material, as well as its degradation and the residual life of critical parts and assemblies.

On the whole, there is practically no data in the literature on the analysis of materials with different types of crystalline structures with an indication of the impact toughness of metals for a wide range of temperatures and the temperature of the viscous-brittle transition.

So, this work tested metals with a different type of crystal structure and revealing the temperature of a viscous-brittle transition

Materials and research methods

To conduct research and identify the features of destruction of materials used for manufacturing critical parts and components of vehicles, metals with various types of crystalline structure BCC (steel 20, steel 45, 09G2S), FCC (12X18H10T and D16) and GPU (VT8) were selected. The materials were subjected to heat treatment according to the modes indicated in Table 1. The studies were carried out on Charpy-type samples with a V-shaped concentrator (working section 8×10, manufactured in accordance with the requirements of GOST 9454–78.

Table 1 – Modes of heat treatment of the studied materials

Material	Type of maintenance
Steel 20	Air hardening 920 °C
Steel 45	Air hardening 870 °C
09G2S	Air hardening 930 °C
12X18H1 0T	Temper 1080 °C
D16	Temper 500 °C, natural ageing
BT8	Double annealing 920 °C 2hr + 570 °C 1hr

The microstructure was studied using a KYENCE-VHX 1000 optical microscope. Facto-graphic analysis was performed using a JSM-3U scanning electron microscope.

Tests for impact bending were carried out on a MK-300 pendulum copra (450 kJ potential energy supply) using the developed at NSTU n.a R.E. Alekseev device for cooling images, the principle of which is described in detail in the literature [8].

To record the temperature during samples cooling, a pt100 sensor with an operating temperature range from –196 to +100 ° C (± 1 ° C) was used.

The total temperature range of the tests was –80 ... + 20 ° C (temperature of climatic cold).

Preparation for testing, testing and processing of the results was carried out according to GOST 9454–78.

Experimental research. Analysis of experimental studies

At Fig. 1 see dependencies of impact strength on test temperature for the studied materials.

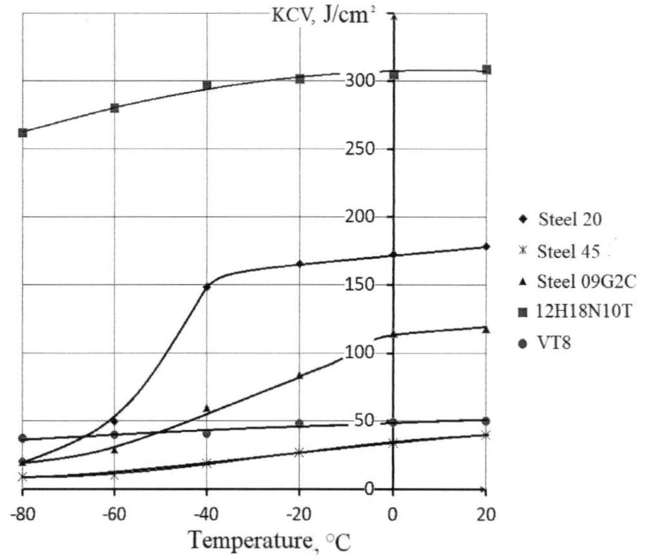

Figure 1. Dependencies of impact strength on test temperature for the studied materials

Fig. 1, does not show the impact strength dependence for the aluminium alloy D16, since the impact strength in the temperature range under consideration does not change and remains 20 J / cm2. As was already noted, to increase the reliability of the operation of the assembly or structure, it is necessary to introduce restrictions on the material by the minimum value of impact strength, which in its turn determines the minimum operating temperature. Most often, the minimum operating temperature is determined by the temperature of the viscous-brittle transition, at which the value of impact strength is significantly reduced, and the likelihood of brittle fracture of the metal also significantly increases. To determine the temperature of the viscous-brittle transition, fracto-graphic studies of fractures of samples were carried out with the determination of the amount of fibrous (B) and brittle (X) components in the fracture. The dependence of the amount of the fibrous component in the fracture of the sample (V,%) on the test temperature for steels 20, 45, and 09Г2C is shown in Fig. 2.

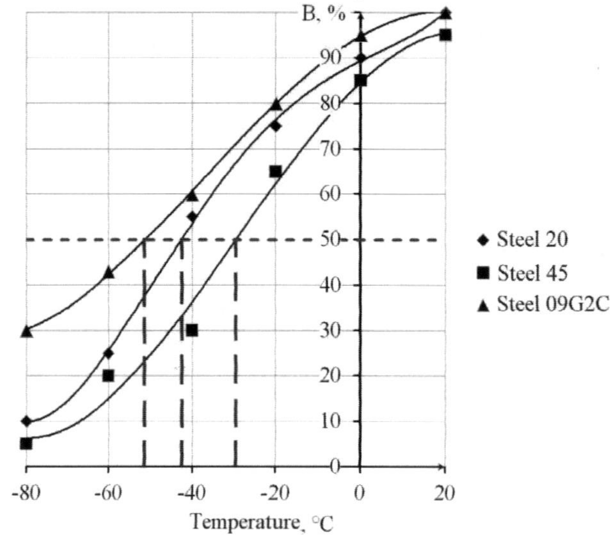

Figure 2. Dependence of the amount of the fibrous component in the fracture of the sample (V,%) on the test temperature for metals with a bcc lattice

An analysis of impact strength dependences (see Fig. 1) as well as the results of fracto-graphic studies of metal fractures (see Fig. 2) shows the following.

1. For metals with a BCC lattice (steel 20, steel 45 and 09Г2С), a significant decrease in impact strength is observed in the temperature range under consideration (5–9 times at t = –80 °C relative to T = +20 °C). The temperature of the viscous-brittle transition according to T_{50} criterion was: for steel 09G2S –28 °C, steel 20 ≈ –43 °C, steel 45 ≈ –43 °C (see Fig. 2).

Thus, metals with a low carbon content (steel 20 and 09G2S) exhibit greater resistance to fracture under impact loads (have higher impact strengths) and have a lower temperature of the viscous-brittle transition. For the metals in question at room temperature, mechanisms of predominantly viscous (pitted) fracture are realized (B = 95 ... 100%). With some decrease in the fracture, facets of cleavage and inter-granular fracture are found.

2. For metals with an FCC lattice (12x18N10T steel and alloy D16) a less intense drop in impact strength is observed with decreasing temperature relative to metals with a bcc lattice (see Fig. 1). For example, for aluminium alloy D16, the level of impact strength does not change with decreasing temperature, but it is at a relatively low level- 20 J / cm2. Steel 12X18H10T significantly surpasses all investigated materials in terms of impact strength. When the temperature drops to temperatures of climatic cold (t = –60 ° C), the level of impact strength remains at the level of 280 J / cm° C, which is explained by its high nickel content. The fracture shape of steel 12X18H10T is strongly distorted, especially at t = +20 ° C, which indicates a high ductility of this material, which remains at a high level even at a temperature of –80 ° C .

3. The VT8 titanium alloy (HCP lattice) is almost 2 times higher than steel 45 and alloy D16 in terms of impact strength, which, as can be seen from the dependence (see Fig. 1), changes with lower intensity. VT8 has a significant reserve of impact strength at a temperature of t = –60 ° C, which remains at the level of KCV = 40 J / cm2.

As shown by previous studies [9, 10], it is important, in addition to changing the toughness, to take into account the change in the fatigue characteristics of metals with decreasing temperature, which are associated with the metal crystal lattice parameter.

Program development to select cold-resistant materials.

To solve the question of what material to choose for manufacturing a part, assembly, or a structure it is necessary to determine a set of properties necessary for its work - mechanical, physico-chemical and, what is also important- technological.

Based on the analysis carried out in the work, we can conclude that improving the reliability of machines and structures, is mostly dependent on the right choice and use of cold-resistant materials. As was shown above, a large number of cold-resistant materials has been developed at present; each of them has its own field of rational use. Despite their great diversity, only a small part of them has the whole range of properties necessary for the normal functioning of technical systems in the Arctic.

The NSTU has developed a program that allows a comprehensive approach to select materials for vehicle assemblies and structures when working in the Arctic and the Far North. This program takes into account mechanical characteristics change within the operating temperature range, as well as the necessary technological properties of the material. The program received a certificate of registration of a computer program No. 2016811331 "Program for selection cold-resistant materials."

Thus, for the competent choice of the right material, the designer must take into account the change in mechanical characteristics at operating temperatures, know the temperature of the viscous-brittle transition, and besides take into account the impact strength index as one of the most important for choosing a material working at low temperatures.

This work was carried out as part of the Russian Science Foundation grant 19-19-00332 "Development of science-based approaches , hardware and software for monitoring damage to structural materials based on artificial intelligence to ensure safe operation of technical facilities in the Arctic"

References

[1] Larionov, V.P. Cold resistance of materials and structural elements: results and prospects / V.P. Larionov, V.R. Kuzmin, O.I. Sleptsov et al. - Novosibirsk: Nauka, 2005 .-- 290 p.
[2] Broek, D. Fundamentals of destruction mechanics / D. Broek.- M .: Vyshhaya shkola, 1980. - 368
[3] Parton, V.Z. The mechanics of destruction / V.Z. Parton.- M., Nauka, 1990 .-- 240 p.
[4] Gordeeva, T.A. Analysis of fractures in assessing the reliability of materials / T.A. Gordeeva, I.P. Zhegin.-.: Mashinostroyenie, 1978.- 200 p.

[5] Sarrak, V.I. Fragile fracture of metals / V.I. Sarrak // Uspekhi Fizicheskikh Nauk .- 1959.- vol. 27.- iss.2.- p.339-361

[6] Thermodynamics: Manual Mashinostroyenie,, 1984. -592 p.

[7] Novitsky, L.A. Thermophysical properties of materials at low temperatures. Directory. / L.A. Novitsky, I.G. Kozhevnikov. - : Mashinostroyenie,, 1975.- 206 p.

[8] Kabaldin, Yu.G. Assessment and diagnostics of metal structural stability at low temperatures based on approaches of nonlinear dynamics and artificial intelligence: monograph / Yu.G. Kabaldinet al. .; under the editorship of Yu.G.. Kabaldin; Nizhny Novgorod State Tech.Un . n.a. R.E. Alekseev - Nizhny Novgorod, 2017 - 210 p.

[9] Kabaldin Y.G., Anosov M.S., Zhelonkin M.V. Russ. Engin. Res. (2018) 38: 848. https://doi.org/10.3103/S1068798X18110047.

[10] Kabaldin Y.G., Anosov M.S., Shatagin D.A. Russ. Engin. Res. (2017) 37: 866. https://doi.org/10.3103/S1068798X17100124.

IOP Conf. Series: Materials Science and Engineering **695** (2019) 012026 doi:10.1088/1757-899X/695/1/012026

Development of design project of mini cross-country vehicle KORSAK

A.G.Kapustin, A.V.Papunin, V.S.Makarov, V.V. Belyakov and M.E. Gnenik

Nizhny Novgorod State Technical University n.a. R.E. Alekseev

E-mail: kapustinalexandr2017@yandex.ru

Abstract. This article reviews different design options for small-size cross-country vehicle KORSAK. That includes 5 options of the vehicle exterior design and the final version, 5 options of the vehicle interior design, 2 options of seats arrangement and the final approved option of the vehicle chassis. The vehicle model made to scale 1:10 is shown in the picture.

Introduction

KORSAK is a ground transport vehicle of improved cross-country capacity designed for off-road mobility.

KORSAK refers to the light cross-country vehicle category. The advantage of this category of vehicles is lower operation costs and capability to move freely over rough terrain areas where even heavy cross-country vehicles could not drive through.

This cross-country vehicle features light weight therefore its resting base on soil is small (fig.1).

Figure 1. General view of KORSAK cross-country vehicles before design concept development

Area of application of the vehicles of interest:

– service & maintenance, inspection and repairs of line structures (pipeline routes, power transmission and communication lines);

– technical diagnostics at any field facility using instruments and equipment up to 200 kg;

– search-and-rescue operations and evacuation activities in natural disaster areas.

As of today, the chassis of KORSAK light cross-country vehicle is developed, the vehicle ground contact configuration and floatation capacity are tested.

Project objective

- The vehicle body shall be designed with new exterior and interior styles matching the existing vehicle body shapes.

Project description

IOP Conf. Series: Materials Science and Engineering **695** (2019) 012026 doi:10.1088/1757-899X/695/1/012026

At work stage one exterior and interior designs of the multi-axle vehicle with improved cross-country capacity were developed. The vehicle model was built using Autodesk inventor and DS CATIA software. Several options of the cross-country vehicle exterior design were developed. These different options are shown in figure 2.

Figure 2. 3D model of KORSAK cross-country vehicle, interior design options.
a - option 2, b - option 3, c - option 4, d - option 5

The main challenge in the development of the cross-country vehicle concept was that not all designer's ideas could be implemented such as shaped hood, arrow-shaped body with non-standard position of side rear-view mirrors, LED lights. A compromise was reached by combining straight and curved lines, and also a smart set of colours. Figure 3 shows the final option of the vehicle exterior design.

Figure 3. Option 6 of final exterior design of KORSAK cross-country vehicle

The key feature of this design is that the cross-country vehicle is covered with black film glued to the vehicle body. This design solution makes the vehicle style more aggressive. The radiator grill of the vehicle is wide to highlight the cross-country nature of the project (fig. 4). The vehicle roof is detachable. The top of the vehicle body is made of fibre glass.

Also, several options of the cross-country vehicle exterior design were developed.

Figure 4. Option 1 of 3D model of KORSAK cross-country vehicle interior design

This option features the dashboard in the form of three wells (fuel consumption, engine temperature, engine RPM and vehicle speed). It is flanked with the audio system camouflaged as deflectors (fig. 5).

IOP Conf. Series: Materials Science and Engineering **695** (2019) 012026 doi:10.1088/1757-899X/695/1/012026

a

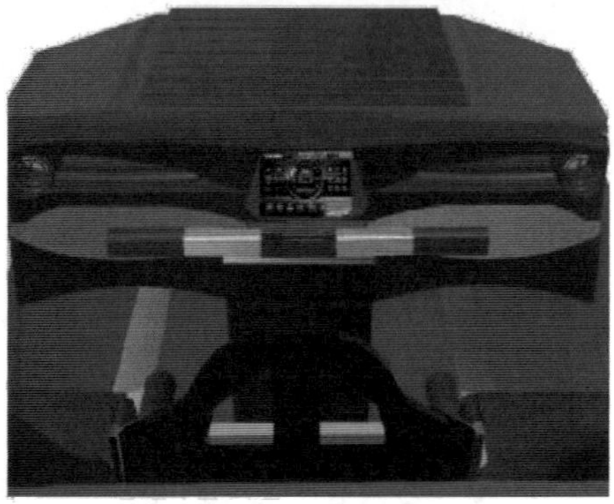

b

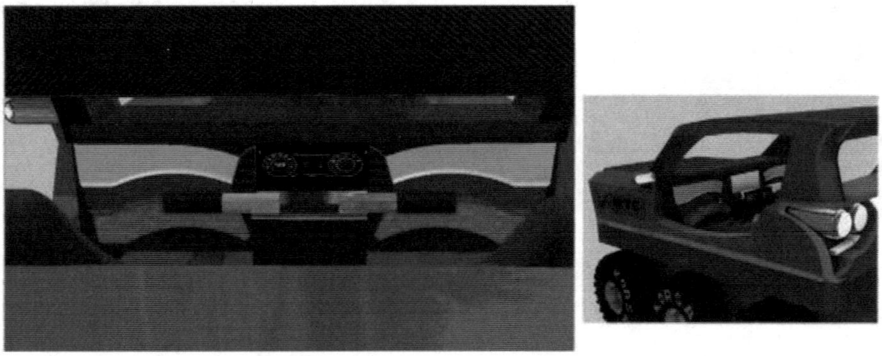

c

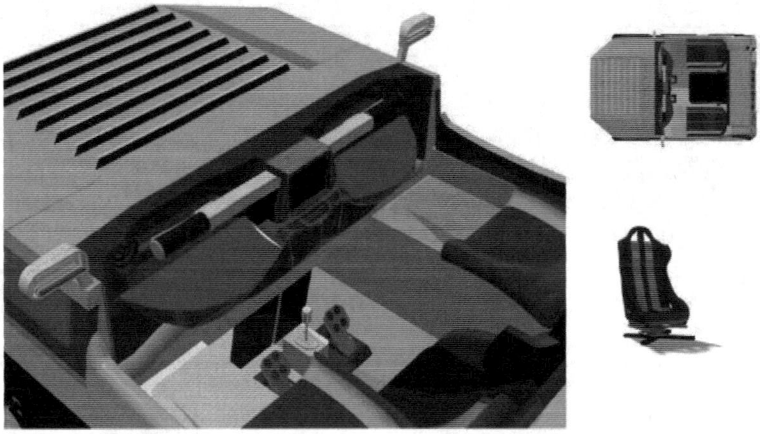

d

Figure 5. 3D model of KORSAK cross-country vehicle, interior design options.
a - option 2, b - option 3, c - option 4, d - option 5

The specific feature of these samples is that a multifunctional tablet with external device connectivity is attached in the centre in place of the dashboard. Deflectors supplying warm air are located near the tablet screen and oriented towards the steering wheel. This solution is practical for cold season. Two extra seats are installed on the left and right sides for comfort and convenience of two passengers. The cross-country vehicle is only 2.5 m long, however, two extra passenger seats can also be accommodated in KORSAK rear side on account of transformable baggage door (fig. 6,7).

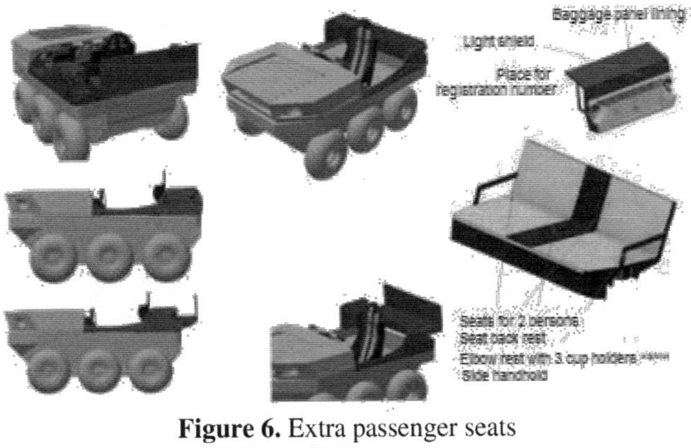

Figure 6. Extra passenger seats

Figure 7. Finished model of cross-country vehicle made to scale 1:10

Conclusions

3D models of different design options of KORSAK wheel-mounted cross-country vehicle were developed.

The most comfortable and technology-savvy design with new appearance was selected for the multi-wheel transport vehicle.

The vehicle model was built to scale 1:10.

Our next step would be the vehicle prototype made to scale 1:1.

References

[1] Papunin A.V. Calculation of the width of the ditch to be overcome by a multi-axle wheeled vehicle / Papunin A.V., Belyakov V.V., Makarov V.S. / Innovations and development prospects of mining engineering and electromechanics: IPDME-2018: Abstracts / St. Petersburg Mining University, St. Petersburg, 2018, S. 35

[2] Papunin A.V. A dynamic model of unsupported pit traversal by a vehicle with 6x6 wheel arrangement / Papunin A.V., Belyakov V.V., Makarov V.S. , Anikin A.A., Vahidov U.Sh. // Paper presented at the IOP Conference Series: Materials Science and Engineering, 2018

[3] Zezyulin D.V. Vehicle "KORSAK" for servicing linear objects / Zezyulin D.V.. Makarov V.S.. Fedorenko A.V.. Belyayev A.M.. Belyakov V.V.//«Works of NNTU» – N.Novgorod. 2014. №4(106) S. 336-341

[4] Papunin A.V., Makarov V.S., Belyaev A.M., Belyakov V.V. Field research of profile trafficability of 6x6 wheel assembly KORSAK vehicle / IOP Conf. Series: Journal of Physics: Conf. Series 1177 (2019) 012051

[5] Papunin A.V., Belyakov V.V., Makarov V.S. Calculation of trench width to be overcome by multi-axis wheeled vehicle / IOP Conf. Series: Earth and Environmental Science 194 (2018) 062027

Sound Generation under Rotary-Screw Propulsion Unit Base Cylinder Bending

I.A. Erasov[1], I.G. Kuklina[1], Y.I. Molev[1], D.S. Mokerov[2] and T.G. Skrebneva[1]

[1] Nizhny Novgorod State Technical University after R.E. Alekseev, 603950m N.Novgorod, Minina Str. 24
[2] Kozma Minin Nizhny Novgorod State Pedagogical University, 603005, N.Novgorod, Ulyanova Str. 1

E-mail: erasoff2013@yandex.ru[a]

Abstract: Theoretical results of sound generation parameters under rotary-screw propulsion unit base cylinder bending are dealt with in this article. Base cylinder parameters impact such as diameter, wall thickness, length as well as such vehicle movement parameters impact as speed and mass on noise level appearing at rotary-screw propulsion unit passing a single obstacle are analyzed in the paper. Sound vibrations onset conditions are shown. Research results allow one to conclude that a considerable sound level generated by rotary-screw propulsion units can appear only when the vehicle mass to rotor diameter ratio is greater than 25. The model produced and the results obtained might be useful when designing cross-country vehicles.

Key words: rotary-screw propulsion unit, bending, noise level

Introduction

Cross-country vehicles providing transport accessibility for all Russia's regions are the only possible way of underpopulated regions development where road building and maintenance is economically impractical. Herewith, one of the basic problems of cross-country vehicles is a high-level noise impact on the operator and passengers which appears when vehicles and road machinery are moving over uneven or rough terrain. Thus, modelling noise level generated by different types of propulsion units in general and especially by rotary-screw propulsion unit, as well as finding ways to lower the noise level are acute research tasks.

Previous papers review

Nowadays only rotary-screw propulsion unit is considered relevant in transporting people and cargoes over water surface and frozen-over rivers [1,2,3,4,5,6,7]. However, due to the fact that not a single vehicle of this type has been produced in series up to now, a number of issues arise when making such vehicles, one of them being noise generation of this or that type.

Methods

The existing models of rotary-screw vehicles movement suggest no rotor base cylinder distortion [1,2,3,4,5,6]. Meanwhile, in operation and forcing an obstacle, the object is impacted by considerable forces (before the vehicle total mass) due to which the rotor will be bent and the vehicle forward movement will turn this bend into a wave process. Beam bending energy is conventionally described by equation [5,6,7,8,9]:

$$W = \frac{m_P \omega^2 A_m^2}{2} \tag{1}$$

where m_P – rotary mass, A_m – bending amplitude, a ω - oscillation frequency equal to the vehicle travel speed V (m/s) to deflection length ratio l (m). At this the maximum deflection length is limited by the base cylinder length. As soon as work is equal to the difference of energy conversion, base cylinder bending capacity can be presented as follows:

$$N = \frac{A}{t} = \frac{W - W_0}{t} = \frac{m_P V^2 (A_m^2 - A_0^2)}{2l^2 \frac{l}{V}} = \frac{m_P V^3 (A_m^2 - A_0^2)}{2l^3} \quad (2)$$

Thus, to determine rotor bending capacity it is necessary to specify the deflection amplitude under a given loading pattern. In this case transverse load causes rotor shaft bending in the plane of load with rotor cross-sections displacement. The bent beam axis is called defected axis or deflection curve. New positions of cross-sections are characterized by linear and angular displacements shown in Fig. 1.

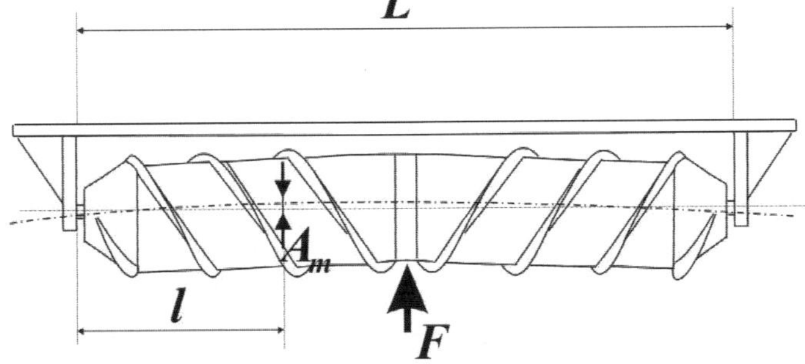

Figure 1. Calculation model to determine rotor bowing value

Deflection curve approximate differential equation has the form of [5,6,9,16,17,18,19]:

$$EIA_m^{//} = M \quad (3)$$

where EI – rotor cross-section rigidity under bending, $A_m^{//}$ - the deflection value second derivative, M – bending moment value, equal to:

$$M = F(1 - \frac{l}{L}) \cdot l \qquad A_m^{//} = \frac{F(1 - \frac{l}{L}) \cdot l}{EI} \quad (4)$$

Substituting l as part of L, that is $l = \varpi L$ we obtain:

$$A_m^{//} = \frac{F L(\varpi - \varpi^2)}{EI} \quad (5)$$

where L- total rotor length, l – a span between force application point and front rotor support. Integrating equation 5 rotor shaft rotation angle equation is obtained first time:

$$\theta = \int \frac{F L(\varpi - \varpi^2) \cdot d\varpi}{EI} = \frac{F L}{EI}\left(\int \varpi d\varpi - \int \varpi^2 d\varpi\right) = \frac{F L}{EI}\left(\frac{\varpi^2}{2} - \frac{\varpi^3}{3} + C\right) \quad (6)$$

taking into account, that in case the load is applied to point $l = L/2$ exactly in the middle of the rotor, angle θ must be equal to 0 we obtain:

$$\theta = A_m^{/} = \frac{FL}{12EI}(6\varpi^2 - 4\varpi^3 - 1) \quad (7)$$

integrating equation 7 for the second time, considering that at the rotor fixturing point (at points $\varpi = 1$ *u* $\varpi = 0$) axle travel A_m must be equal to 0, the required rotor shaft deflection equation is obtained:

$$A_m = \frac{FL}{12EI}(2\varpi^3 - \varpi^4 - \varpi) \quad (8)$$

Simultaneously solving equations 2 and 8, considering that a single rotor maximum load is equal to a half of total vehicle mass we obtain:

$$N = \frac{m_P V^3 \left(\frac{FL\varpi}{12EI}(2\varpi^2 - \varpi^3 - 1)\right)^2}{2L^3 \varpi^3} = \frac{m_P V^3 G^2 (2\varpi^2 - \varpi^3 - 1)^2}{1152 EIL\varpi} \quad (9)$$

Taking into account that rotor resisting moment is conventionally estimated by value [5,6,9]:

$$I = \frac{D^4(1 - C^4)}{2}, \quad где \;\; C = \frac{d}{D} = 1 - \frac{\delta}{D} \quad (10)$$

where D – base cylinder diameter, δ - wall thickness. Taking into account that rotary mass is equal to its volume multiplied by material density $m_P = \rho L \pi D \delta$ the maximum radiated sound power appearing at the base cylinder bending is determined by equation:

$$L_P = 10\lg\left(\frac{\rho \delta V^3 G^2}{2350 ED^3 (1 - C^4)}\right) \quad (11)$$

Solution of equation 14 is shown in Fig. 2

Results

The results obtained indicate that for actual vehicles with the mass of over 2 tons and the base cylinder diameter 0,3 m, the generated noise level under rotor bending will not exceed 5 dB. For light cross-country vehicles with the mass of up to 1 ton, base cylinders with diameter of over 0,2 m completely exclude noise generation under bending. Thus, the necessary dependences have been obtained to design rotary-screw propulsion units of cross-country vehicles completely meeting normative noise requirements [11,12,13,14,15].

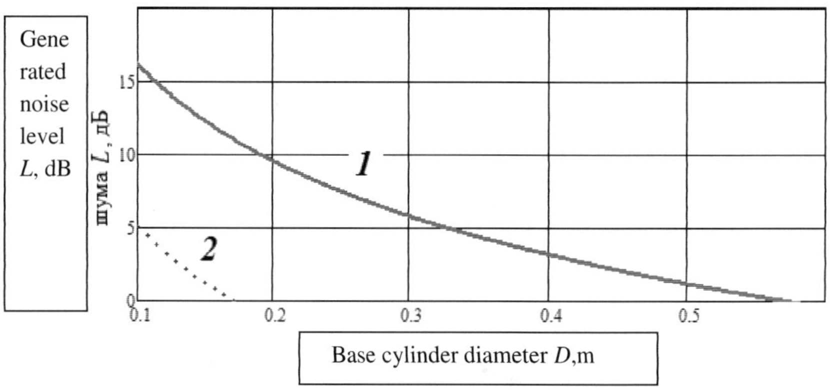

Figure 2. Generated noise level change, appearing under base cylinder bending, of its diameter;
1- For 2,5 ton vehicle; 2 – for 750 kg vehicle

Acknowledgement

This work was carried out at the NNSTU named after R.E. Alekseev, with financial support from the government in the face of the Russian Ministry of Education under the agreement 14.577.21.0222 of 03.10.2016. The unique identifier of the project: RFMEFI57716X0222. Theme: "Development an experimental model of an amphibious autonomous transport-technological complex with an intelligent control and navigation system for year-round exploration and drilling on the Arctic shelf".

References

[1] Krasheninnikov M., Kulashov A., Shapkin V., Koshurina A. The concept and methodology of creating the universal life-saver with rotary-screw mover // Lecture Notes in Electrical Engineering. 2013. – T. 195 LNEE. – № VOL. 7 «Proceedings of the FISITA 2012 World Automotive Congress».

[2] Krasheninnikov M., Koshurina A., Vasilyev I., Smirnova E. Use mobile robots groups for rescue missions in extreme climatic conditions // Procedia Engineering, Volume 100, 2015, Pages 1242–1246 doi: 10.1016/j.proeng.2015.01.489.. 477 – 490

[3] Korushova J.V., Krasheninnikov M.S., Koshurina A.A., Lyakhmanov, D.A. Multifunctional mobile complex of transport-technological machines for the Arctic // IOP Conference Series: Materials Science and Engineering 2018 IOP Conf. Ser.: Mater. Sci. Eng. 393 012119

[4] Koshurina A.A., Krasheninnikov M.S., Dorofeev R.A. Strength Calculation and Analysis of Equalizer Beam Embodiments for the Operated Equalizing Beam Suspension of the Universal Rotor-screw Rescue Vehicle for the Arctic// Procedia Engineering, Volume 150, 2016, Pages 1263-1269

[5] Ice loads dynamics for model scale cylinders of various diameters Zvyagin, P., Dobrodeev, A., Sazonov, K. 2017 Proceedings of the International Conference on Port and Ocean Engineering under Arctic Conditions, POAC

[6] Kulyashov A.P., Kuklina I.G. Oscillations estimate of a rotary-screw vehicle considering its suspension parameters and propulsion units-and-soil linear contact // Journal «Construction and road machinery». 2011. № 7. P. 45.

[7] Transport and technological vehicles roadbed: college textbook[eds. V.V. Belyakov, A.A. Kurkin]; NNSTU n.a. R.E.Alekseev – N.Novgorod, 2014. – 447p.

[8] Cross-country transport and technological vehicles. Fundamentals of the theory of motion /under the general editorship of V.V. Belyakov and A.P. Kulyashov – N.Novgorod: publishing house ТАЛАМ, 2004. -961p.

[9] Vakhidov U.Sh., Kitov A.G., Sogin A.V., Shapkin V.A., Shapkina Y.V. Noise and vibration parameters of transport and technological vehicles // International Journal of applied and fundamental research. 2014. № 7. P. 8-11.

[10] Dynamics, critical speeds and balancing of thermoelastic rotors Yeliseyev, V.V. 2015 Lecture Notes in Mechanical Engineering 22, c. 129-136

[11] ISO 2041:2009 Mechanical vibration, shock and condition monitoring — Vocabulary - Third Edition

[12] Mechanical vibration - Testing of mobile machinery in order to determine the vibration emission value; German version EN 1032:2003

[13] ISO 2631-1:1997 Mechanical vibration and shock — Evaluation of human exposure to whole-body vibration — Part 1: General requirements (MOD).

[14] ISO 5349-1:2001 Vibration. Measurement and evaluation of human exposure to hand-transmitted vibration. Part 1. General requirements

[15] ISO 5349-2:2001 Vibration. Measurement and evaluation of human exposure to hand-transmitted vibration. Part 2.Requirements for measurement at the workplace/

[16] V Klubnichkin, E Klubnichkin, G Kotiev, S Beketov and V Makarov Interaction between elements of the track ground contacting area with the soil at curvilinear motion of the timber harvesting machine/ IOP Conference Series: Materials Science and Engineering Volume 386, 2018, Pages 012016

[17] A Papunin,V Belyakov,V Makarov,A Anikin and U Vahidov A dynamic model of unsupported pit traversal by a vehicle with 6×6 wheel arrangement/ IOP Conference Series: Materials Science and Engineering Volume 386, 2018, Pages 012001

[18] E Klubnichkin,V Klubnichkin and G Kotiev Theoretical research of soil packing by timber harvester running gear IOP Conference Series: Materials Science and Engineering Volume 386, 2018, Pages 012025

[19] A Lipin, Y Molev, D Mokerov, A. Strizhak,VKhudyakov Ways of decreasing noise impact on operator by changing ro-tary-screw propulsion units natural frequency of vibration IOP Conference Series: Journal of Physics: Conf. Series 1177 (2019) Pages 012040

[20] I Erasov, Yu Molev, D Mokerov, T Shetulova Determining friction coefficient between the base cylinder of rotary screw propulsion unit and various ice types in different daylight surface IOP Conf. Series: Materials Science and Engineering 386 (2018) 012006
[21] Y Molev, D Mokerov, S Ivanov, M Saushkina, Y Palutin Sound power spectra modelling of the vehicle in motion equipped with rotary-screw propulsion unitIOP Conf. Series: Journal of Physics: Conf. Series 1177 (2019) 012034

IOP Conf. Series: Materials Science and Engineering **695** (2019) 012029 doi:10.1088/1757-899X/695/1/012029

Improvement of car repairability based on the material destruction assessment during its operation

N.A. Kuzmin, D.A.Goncharova, G.V. Pachurin, A.A. Filippov and I.A. Korotunova

Russia, Nizhniy Novgorod, Nizhniy Novgorod State Technical University named after R.E. Alekseev

E-mail: kuznntu@mail.ru

Abstract. Parts and structural elements of vehicles during operation are affected by cyclic loads and various temperatures. At the same time, more than 80% of all types of metal structures destruction are fatigue-like, which leads to huge financial losses, and often to fatal accident. Therefore, issues of improving the reliability and safe operation of automobile parts subjected to cyclic loads in air at different temperatures are important in modern industrial science. That's why, it is necessary to know the laws of fatigue behavior of structural materials obtained by different types and modes of technological processing in specific operating conditions in order to predict their operational reliability increase.

Changes in the metal during fatigue are directly reflected on the curves of changes in the samples current deflection, since they reflect the antogonism of two mutually opposite processes - hardening and softening, directly related to the processes that occur in the material structure during fatigue.

This article presents the possible analysis of the fatigue failure kinetics of some test structural materials using the curves of the samples current deflection. The assessment of structural materials destruction allows us to choose structural materials for car parts, taking the conditions of its operation, types and modes of its manufacturing technology into account in order to increase its maintainability and service life.

Keywords: *vehicles, metal parts of a car, reliability, safety, fatigue resistance, materials damageability.*

Introduction

When operating in corrosive environment metal parts and structural elements of a car are subjected to vibration loads, that can lead to their fatigue failure, and consequently to accidents, even those with casualties [17-19]. Fatigue failure of structural materials depends on the material characteristics, its processing procedures and loading conditions (medium and voltage amplitude) [7, 8, 14].

In the corse of manufacturing process a car part is put through various technological processing procedures (thermal, volume and surface plastic, welding, etc.) [9]. Apart from the influence of a large number of internal and external factors [20], the complexity of full-scale testing in different environments [15, 26] complicates the elaboration of reliable criteria for assessing the fracture resistance of metals and alloys under cyclic loads.

In this regard, and taking into account the intensive development of volume plastic deformation advanced technique, the problem of creating fatigue fracture patterns in order to predict and increase the service life of strain-hardened metal materials and, consequently, decrease materials consumption is very important.

The process of fatigue failure of metallic materials is tentatively divided [10, 16] into three stages:

1) the period before the onset of fatigue microcracks;
2) subcritical crack growth;
3) quick rupture

The curves of cyclic hardening (softening) reflecting the processes of structural changes [24] are important when decscribing material behaviour under cyclic loading. Hardening is a preparatory stage in the process of fatigue of metal materials, after which at certain values of alternating stresses for a certain number of cycles, loosening occurs that ends with the fatigue fracture onset and expansion.

All this structural damage, which is the result of hardening-softening antagonism directly related to the processes that occur in the structure of the material during fatigue, affects the sample running deflection in the course of cyclic loading.

The purpose of the work carried out in NSTU named after R.E. Alekseev is to increase the maintainability of a car through establishing patterns of metallic materials destruction during its operation.

The object of research is strain-hardened metallic materials widely used for car elements construction.

The subject of research is the laws of fatigue failure of automotive structural materials of various classes and structure.

Exploratory procedure

In this work fatigue failure of automotive materials was carried out on cylindrical corset-shaped samples made of the same-supply bar stocks for each structural material. The characteristics of the samples are presented in table 1.

Table 1 – Sample materials state and structure

Material	Sample initial state	Material microstructure
Copper M1	Cold-rolled	Grain size 0,04-0,08 мм
Copper M1	Annealing 540°C 2h vacuum 1,33^{-3}10 Pa. Furnace cooling	Grain size 0,04-0,08 мм
Brass Л63Т	Cold-rolled	Grain size 0,04 мм
Aluminium alloy В95пчТ2	Hardening465-475°C (выдержка 1ч), water, correcting (stretching when fresh-hardened 1,7%); aging (12O°C, 5h, 180°C,6h)	–
Steel 20ХI3	Hardening 1030°C, oil, tempering 600 – 640°C.	Martensite
Steel 14Х17Н2	Hardening1030°C, oil, tempering 620 -660°C.	Martensite
Steel 35ХГСА	Hardening900°C, storage 20 min, oil, tempering 425°C	Troostosorbite

The samples were tested using the symmetric cantilever bending scheme with a rotation frequency of 50 Hz on a MIP-8 machine equipped with a phase synchronizer and an optical microscope (x37) with stroboscopic illumination in order to observe structural changes, the occurrence and development of fatigue cracks on the surface of a cylindrical sample with continuous cyclic loading. To provide corrosive medium for testing material a special device to the fatigue machine was designed and manufactured [2, 3].

The initial microstructure of the samples was determined using an AKASHI optical microscope. In addition, dangerous section samples with an etched surface were removed from a fatigue machine for detailed study and photographing once in a certain number of loading cycles.

The outcomes and their consideration

It was shown previously [1, 23] that the curves of changes in the deflection of the sample during the tests reflect the peculiarities of the kinetics of the material fatigue behavior.

Using them, one can determine the moment of a fatigue crack onset leading to the deflection increase, and also evaluate its expansion speed. Such information becomes especially valuable when cyclic loading the materials under conditions other than normal, that is, when direct observation of the sample surface is inconvenient or impossible, for example, in a corrosive environment. Therefore, the MIP-8 fatigue machine was equipped with a dial gauge (accuracy 0.01 mm), which allowed scanning the in the current deflection of the sample during fatigue loading both under normal conditions and in a corrosive environment (3% solution of sea salt in water).

The results of the analysis of the sample deflection changes during fatigue at room temperature demonstrate [4, 5, 21, 25] that the curves have three sections (Fig. 1):

1 - a sharp decrease for annealed or an increase for cold-rolled materials;

2 - stage of the deflection stabilization;

3 - its sharp increase associated with the sweeping destruction of the material. In this case, an increase in the sample deflection after the stabilization stage corresponds to the moment of macrocracks onset ~ 1.0 mm long on its surface (for example, Fig. 1).

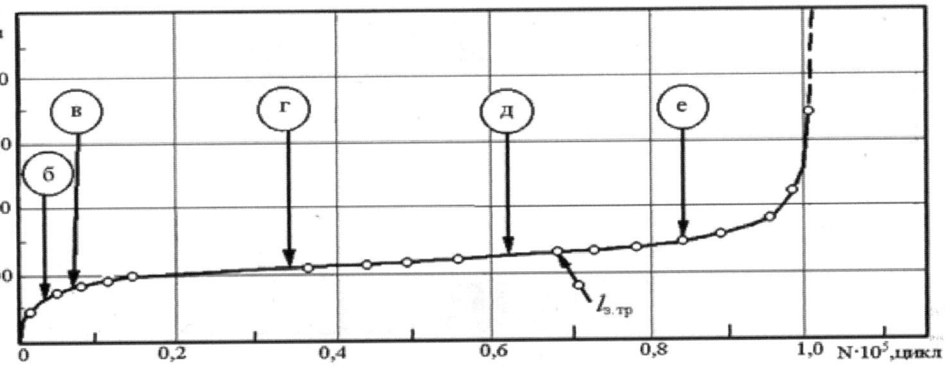

Figure 1. Dynamics of the current deflection of a sample from V95pchT2 aluminum alloy depending on the number of loading cycles at σa = 248 MPa (durability to complete destruction of the material Np = 1.15 · 105 cycles): a) initial state of the material; b) N / Np = 2.4%; c) N / Np = 7.4%; d) N / Np = 14.8%; d) N / Np = 29.3%; e) N / Np = 51.0%; g) N / Np = 73.8%; h) - N / Np = 88.0%

Figure 2 shows structural changes in the surface of a V95pchT2 aluminum alloy sample depending on the number of loading cycles at σa = 248 MPa. The letter designations correspond to those on the curve of the current deflection of Fig.1.

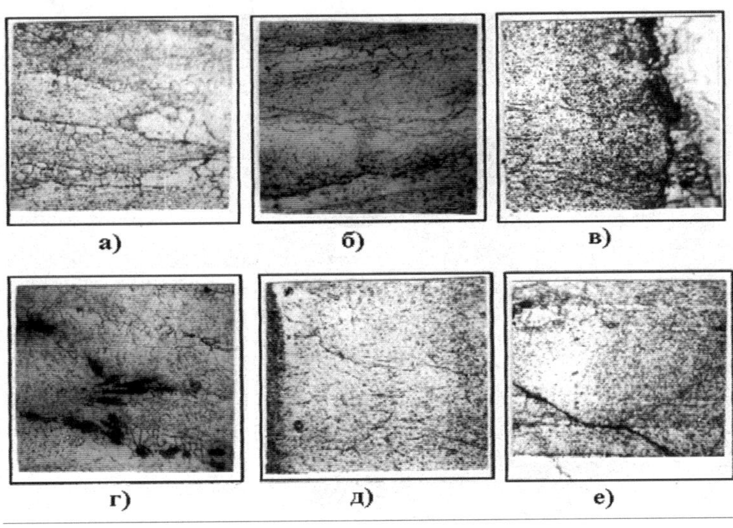

Figure 2. Kinetics of the destruction of the V95pchT2 aluminum alloy depending on the number of loading cycles at σa = 248 MPa (durability to complete destruction of the material Np = 1.15 · 105 cycles): a) initial state of the material; b) N / Np = 2.4%; c) N / Np = 7.4%; d) N / Np = 14.8%; d) N / Np = 29.3%; f) N / Np = 51.0%; g) N / Np = 73.8%; h) - N / Np = 88.0%; x 450

Figure 3 shows a developed fatigue crack on the surface of V95pchT2 aluminum alloy sample.

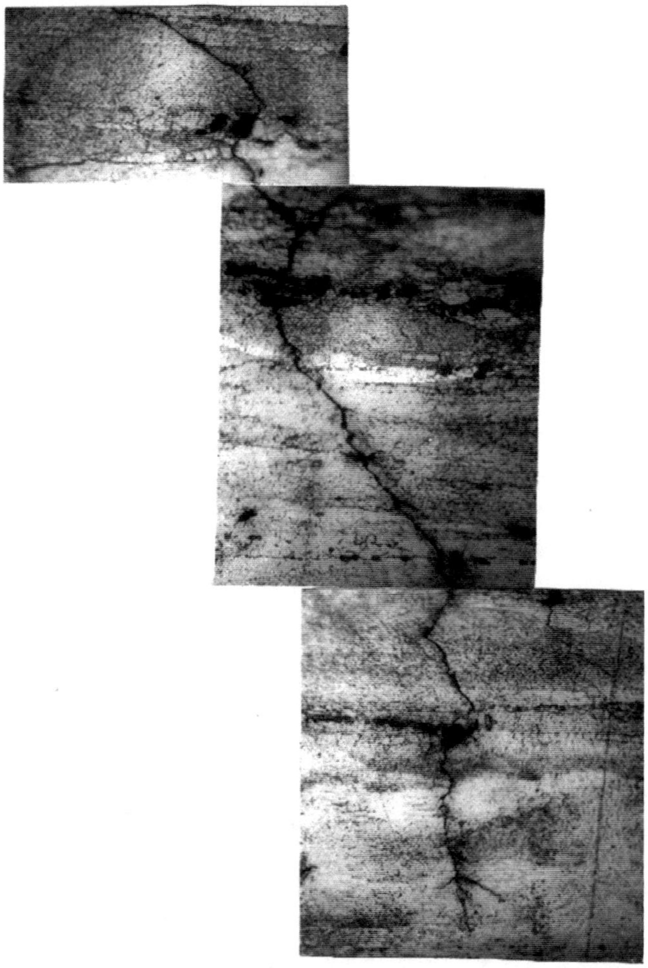

Figure 3. A developed fatigue crack on the surface of V95pchT2 aluminum alloy; x450

It is known [11, 12] that corrosion-fatigue fracture of metals and alloys is initiated by surface damage, erosion, cavities, intergranular corrosion on it and ,therefore, develops, as a rule, from several sites. This determines the multi-site nature of the corrosion-fatigue fracture development.

However, despite the distinctive features of the corrosion-fatigue destruction process, the curves of changes in the samples current deflection under cyclic loading in a 3% sea salt solution have a qualitatively the same form as when tested in air [6, 13].

This is due to the fact that, along with the mechanisms of hardening and softening, a decrease in the clear section of the sample is the determining factor for obtaining qualitative data on the current state of materials during cyclic loading according to the parameter of the current deflection change. Therefore, despite the fact that it is rather difficult to distinguish clearly the process of corrosion-fatigue failure into a number of periods from a physical point of view, as the case when tested in air, nevertheless, such conventional division can be justified to facilitate the study of the fatigue failure kinetics. Some other authors hold the same opinion [22].

This approach allows us to divide tentatively the fatigue-and-corrosive destruction process into three distinctive phases:

- the incubation phase, which is characterized by processes associated with the adsorption of the medium on the surface of metal or oxide films through selective anodic dissolution, hydrogenation of the cathode sections and other processes induced by cyclic mechanical stresses. These processes lead to the formation of pitting or microcracks with a depth sufficient for a noticeable concentration of mechanical stresses;

- the phase of corrosion-fatigue crack growth to critical sizes;
- accelerated rupture phase.

During the third phase corrosive medium has practically no effect on the nature and duration of the metallic material destruction, however, the duration of the first and the second phases is significantly affected not only by corrosive medium but also by the material properties and its structural state caused by the technological treatment and loading characteristics (the amplitude of the applied voltage and cycle duration).

The analysis of the obtained results deminstrates the dependence of the qualitative change in the samples current deflection from their live section alternation during cyclic softening,as well as in air under heavy loads when a fatigue crack originates around the entire perimeter of the dangerous section of the sample (the so-called "circular" crack onset and growth).

Conclusion

Thus, the current deflection curves in addition to metallographic, fractographic and other methods of kinetics analysis of the fatigue failure are a very important integral characteristic of the processes occurring under metal materials fatigue loading.

The assessment of structural materials destruction allows us to choose structural materials for car parts, taking the conditions of its operation, types and modes of its manufacturing technology into account in order to increase its maintainability and service life.

Bibliography

[1] **Kruchinin VV, Sofronov Yu.D.** Studying the propagation velocity of fatigue cracks by measuring the deflection of a sample. // Strength of metals under cyclic loads: Sat. - M .: Nauka, 1967. - S. 107-117.

[2] **Pachurin G.V., Guslyakova G.P.** Device for testing fatigue when bending a rotating sample. Copyright certificate No. 920456, Bulletin of inventions, 1982, No. 14, - S. 158.

[3] **Pachurin G.V., Guslyakova G.P.** Chamber for fatigue testing of samples according to the cantilever bending scheme with rotation in the medium of solid refrigerant // Factory Laboratory. - 1987, No. 1. - S. 87-88.

[4] **Pachurin G.V.** Kinetics of fatigue failure of copper M1 and brass L63 // Non-ferrous metallurgy. Proceedings of the universities of the USSR. 1989. No. 1. - S. 96-101.

[5] **Pachurin G.V.** Fatigue failure at normal temperature of pre-deformed alloys. // Metallurgy and heat treatment of metals. 1990. No. 10. - S. 35-38.

[6] **Pachurin G.V., Guslyakova, G.P., Vlasov, V.A.** Cyclic durability of pre-deformed steels 20X13 and 14X17H2 // Izv. Universities. Ferrous metallurgy. 1991. No. 5. - S. 33-35.

[7] **Pachurin G.V., Guslyakova G.P.** Optimization of processing conditions in order to increase resistance to corrosion-fatigue fracture of metallic materials. - N. Novgorod: VSNTO Mashprom, 1991. - 72 p.

[8] **Pachurin G.V.** Improving the corrosion durability and operational reliability of products from strain-hardened metal materials: textbook. manual for university students / NSTU. - N. Novgorod, 2005. - 132 p.

[9] **Pachurin G.V., Gushchin A.N., Galkin V.V., Pachurin V.G.** The theoretical basis for increasing the operational durability of stamped metal products: textbook. manual for university students / NSTU. - N. Novgorod, 2006. - 173 p.

[10] **Pachurin, G.V.** Cyclic corrosion durability of deformed structural materials // Metal Technology. 2003. No. 10. - S. 16-21.

[11] **Pachurin G.V.** The durability of stamped structural materials in air and in a corrosive environment // Procurement in mechanical engineering. 2003. No. 10. - S. 21-27.

[12] **Pachurin G.V.** The durability of sheet stamped materials in air and in a corrosive environment // Materials. - 2003. No. 7. - S. 29-32.

[13] **Pachurin G.V.** Theoretical Foundations of Increasing Corrosion Durability of Hardened Metals and Alloys // Successes in Modern Natural Sciences. - 2011. - No. 2. - S. 95-99.

IOP Conf. Series: Materials Science and Engineering **695** (2019) 012029 doi:10.1088/1757-899X/695/1/012029

[14] **Pachurin G.V.** Corrosion Durability of Deformation-Hardened Metal and Alloy Products: A Training Manual. - 2nd ed., Ext. - St. Petersburg: Publishing House "Lan", 2014. - 160 p.

[15] **Pachurin G.V.** Fatigue resistance at different temperatures of annealed and hardened brass L63 // International Journal of Experimental Education. - 2014. - No. 3-1. - S. 119-126.

[16] **Pachurin G.V.** To the question of hypotheses of corrosion-fatigue fracture of metals and alloys // Fundamental Research. - 2014. - No. 3-1. - S. 28-34.

[17] **Pachurin G.V., Kudryavtsev S.M., Soloviev D.V., Naumov V.I.** The body of a modern car: materials, design and production: Textbook / Ed. G.V. Pachurina. - 3rd ed., Revised. and add. - St. Petersburg: Publishing House "Lan", 2016. - 316 p.

[18] **Pachurin G.V., Mindrin V.I., Filippov A.A.** Safety of operation of industrial equipment and technological processes: a training manual / G.V. Pachurin, V.I. Mindrin, A.A. Filippov; under the general. ed. G.V. Pachurina. - Stary Oskol: TNT, 2017. - 192 p.

[19] **Pachurin G.V., Galkin V.V., Pachurin V.G.** Durability of hardened metals and alloys: a training manual / G.V. Pachurin, V.V. Galkin, V.G. Pachurin. - Stary Oskol: TNT, 2017. - 228 p.

[20] **Naumov V.I., Pachurin G.V.** The role of impurities and ligatures in ensuring the technological properties of metals, alloys and melts: a training manual / V.I. Naumov, G.V. Pachurin; under the general. ed. G.V. Pachurina. - Stary Oskol: TNT, 2017. - 376 p.

[21] **Pachurin G.V., Goncharova D.A., Kuzmin N.A., Filippov A.A., Nuzhdina T.V.** The process of fatigue damage to sheet automotive steels // Technology of metals. - 2019. - No. 8. - S. 16-22.

[22] **Pokhmursky V.I.** Corrosion fatigue of metals / V.I. Pohmursky. - M .: Metallurgy, 1985 .-- 207 p.

[23] **Segal Ya.S.** Using registration of deflection of a sample for studying the process of fatigue // Strength of metals under cyclic loads: Sat.- M .: Nauka, 1967. - P. 66-71.

[24] **Terentyev V.F.** Fatigue of metallic materials / V.F. Terentyev; Repl. ed. N.P. Lyakishev. - M .: Nauka, 2003. - 254 p.

[25] **Goncharova D.A., Pachurin G.V., Filippov A.A., Kuzmin N.A.** Fatigue destruction of automo-bile construction materials // 3RD SCIENCE AND TECHNOLOGY SEMINAR "MOBILITY OF TRANSPORT AND TECHNOLOGICAL MACHINES", MTTM 2018, Nizhny Novgorod, August 28, 2018 - P. 012025.

[26] **Filippov A.A., Pachurin G.V., Naumov V.I., Kuzmin N.A.** Low-Cost Treatment of Rolled Products Used to Make Long High-Strength Bolts // Metallurgist. - 2016. - Vol. 59. - Nos. 9-10. January. - S. 810-815.

IOP Conf. Series: Materials Science and Engineering **695** (2019) 012030 doi:10.1088/1757-899X/695/1/012030

Optimization of perlite steels mechanical properties for car fasteners stamping

A.N. Kuzmin, A.A. Filippov, G.V. Pachurin, D.A.Goncharova

Russia, Nizhniy Novgorod, Nizhniy Novgorod State Technical University named after R.E. Alekseev

E-mail: pachuringv@mail.ru

Abstract. The car fasteners made of pearlitic steels by cold stamping method are to meet special requirements for their structural and mechanical characteristics.

The hardware is made of hot-rolled steel, which cannot be used straight after the delivery for fasteners manufacturing due to the poor quality of the structure and surface layer. Therefore, calibrated rolled stock gets heat treatment before drawing. In addition, hardware of strength class 8.8 and higher obtained from alloy steels using cold heading machines are subjected to temper hardening, which often leads to surface decarburization, cracking and distortion of finished products, that afflicts labor and energy efficiency, and, thus, the final cost of production.

The article suggests a method for 35X and 38XA pearlitic steels stock treatment excluding recrystallization annealing, rod-like items temper hardening, and including isothermal operation to prevent decarburization, risks and distortion. Passing through the die, hardening the surface due to volume strain and thread rolling impart to M8 bolts and studs the mechanical properties corresponding to strength grades 8.8 and 9.8.

This study also has practical application, it allows manufacturers to determine the appropriate mode for producing automotive fasteners.

Key words: *pearlitic steels, drawing, structure, mechanical properties, cold stamping, automotive fasteners.*

Introduction

Reliable operation of metal products to a great extent depends on metal characteristics [17,20], operating conditions [6,7] and its processing conditions [8,9] that form material structure [1,2,23] provided by the processing conditions refinement[3.21].

Special requirements are imposed on bar bolts manufactured from pearlitic steel [4,22] by cold stamping. The most common and productive method of manufacturing final products with metal parts of high quality is their metal-roll upset having uniformly distributed ductility, structure, manufacturability and no defects [11].

A huge stock of hardware products (bolts, studs and screws) that have no special requirements for strength properties is mainly made of State Standard 1050-2013, State Standard 4543-2016 and State Standard 380-2005 steel . These standards do not regulate metal structure and surface parameters when passing through the draw die and then being upset, that involves risks of such defects as strain cracks [5,14].

The chemical composition and mechanical parameters of metal for cold stamping on press equipment for cars, tractors and other vehicles are regulated by State Standard 10702-2016. The mechanical parameters of high-duty equipment are regulated by the active ISO 898-1: 1999 standard. The required strength class is ensured by the choice of steel grade [15,18] and the heat treatment of rolled products and final products [16].

Raw hot-rolled metal from pearlite-grade steels purchased from suppliers of metallurgical products has perlite-ferrite structure. Heat treatment and drawing of the main products before fitting are to form necessary surface and microstructure of rolled products [10,13], which guarantee high quality of metal products after cold upset [12,19].

By State Standard 1759.4-87, elongated products from cold upset alloy steels alloy steels with strength class 8.8 and higher must be subjected to temper hardening . However, this causes decarburization, cracking and deformation of the surface , afflicts labor and energy efficiency, and, thus, the final cost of production.

The purpose of the work carried out in NSTU named after R.E. Alekseev consists in developing an appropriate rolling preparation method, including isothermal mechanical treatment of pearlite steels 35X and 38XA, and excluding recrystallization annealing and temper hardening from the manufacturing process. This will reduce the risks of distortion of long-measuring automobile fasteners and decarburization of their surface. In this case, the surface and mechanical parameters of the stock must meet State Standard 10702-2016.

The object of this development is hot-rolled steels of pearlite class 35X and 38XA for cold headed automotive metalware.

Subject of study. The mode of technological processing of hot-rolled steels 35X and 38XA for cold headed automotive metalware.

Research Methodology. We studied samples of rolled steels 35X and 38XA, which are widely used for the engine, body, automotive and special equipment fasteners. The chemical composition of the metal studied is shown in table 1.

Table 1 - The chemical composition of the steels 35X and 38XA studied

Steel grade	Content of elements, %					
	C	Mn	Si	P	S	Cr
35X	0,34	0,52	0,19	0,009	0,01	0,87
38XA	0,38	0,59	0,20	0,015	0,009	0,89

The scheme of rolling products isothermomechanical preparation was considered to obtain rod-shaped products with a low-shaped head and various studs of strength class 8.8 or more using cold-stamping automatic machines with hardening and tempering operations excluded. Such stock preparation includes heat treatment of the initial bundle; etching; drawing; isothermal treatment (470 ° C); etching; drawing; compression.

The initial mechanical characteristics of rolled steels 35X and 38XA, supplied by metallurgical company, are presented in table 2.

The scheme of rolled steels 35X and 38XA isothermomechanical preparation is displayed in Fig. 3

The hot-rolled metal 35X and 38XA microstructures were studied and revealed to be"perlite + ferrite". These microstructures can be seen in Fig. 1 and fig. 2

Table 2 - Mechanical properties of the initial stock 35X and 38XA

Steel grade	Stock diameter, mm	Strength and ductility characteristics			
		δ	Ψ	$\sigma_в$,	$\sigma_{0,2}$
		%		MPa	
35X	10,1	21	59,6	724	605
38XA	10,3	20,5	59	750	610

IOP Conf. Series: Materials Science and Engineering **695** (2019) 012030 doi:10.1088/1757-899X/695/1/012030

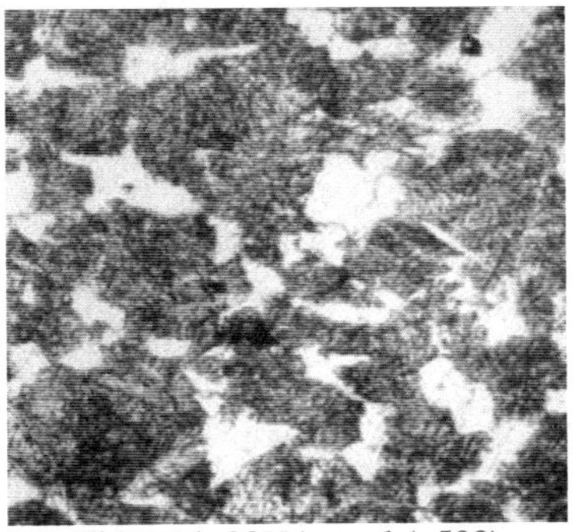
Figure 1. 38XA steel (x500)

Figure 2. 35X steel (x500)

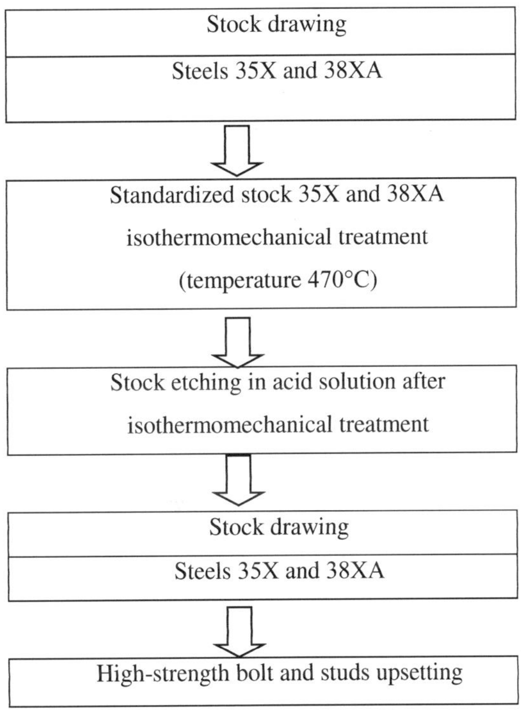

Figure 3. Stock isothermomechanical preparation scheme

The stock microstructure for metalware products with the diameter of 7.8 mm was formed by isothermomechanical treatment (annealing of the 2nd type). It is known that the transformation of austenite in the nitrate bath of these steels occurs in the range of sorbitol transformation of 650–470 ° C. An austenitic eutectoid type is obtained. The considered method of thermal cooling at 470 ° C after drawing is suggested for cold stamping when fitting short, medium and long studs and bolts with massive molded heads. When cooling the wire with the diameter of 8.2–8.6 mm at a temperature of 470 ° C the completion of the austenitic stock transformation takes 3.9 minutes.

Metal samples were got by punching dies with the diameters of 8.1; 8.2; 8.3; 8.4; 8.5 and 8.6 mm, followed by treatment in a nitrate bath (470 ° C). The final size was obtained with the diameter of 7.8 mm. The M8 bolts and studs produced were subjected to rupture in accordance with State Standard 1759.4-87.

The outcomes and their consideration

The chemical composition, strength and ductility properties of 35X and 38XA steels complied with State Standard 10702-16. The microstructure of the stock supplied consists of perlite with ferrite.

After isothermal treatment samples with the diameter of 8.1; 8.2; 8.3; 8.4; 8.5 and 8.6 mm in a salt bath at a temperature of 470 ° C have the structure of perlite and sorbitol, it is shown in Fig. 4 and fig. 5.

Figure 4. 35X steel (x500)

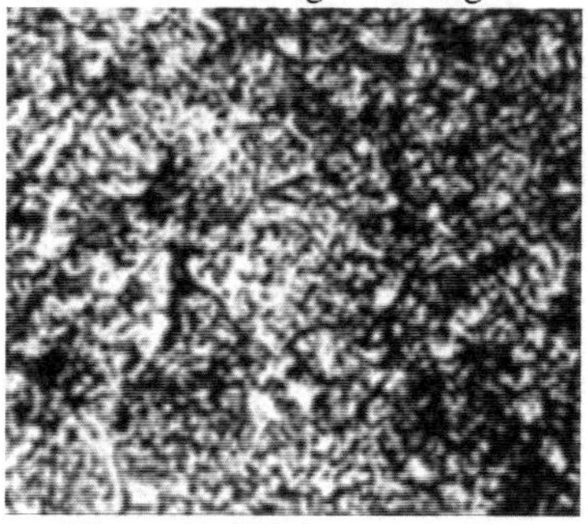

Figure 5. 38XA steel (x500)

The hardness of 35X steel samples after the above mentioned treatment is HB 249, and that of the 38XA steel samples is HB 254.

It has been found out that the strength characteristics of isothermally (470 ° C) treated rolling steel increase monotonously with an increase in deformation when compressed from 4.9 to 17.7%. Thus,when rolling with the compression from 4.9 to 17.7% the tensile strength of 35X steel increases from 838 to 971 MPa, and that of the 38XA steel from 918 to 1067 MPa.

The conventional yield strength also increases after isothermomechanical preparation. It was revealed that the conventional yield strength of rolled 35X steel with compression deformation from 4.9 to 17.7% increases from 769 to 887 MPa, and that of 38XA steel from 797 to 948 MPa.

Rolled stock isothermomechanical treatment at a temperature of 470 ° C followed by drawing with compression from 4.9 to 17.7% provides a monotonic decrease in the relative elongation for 35X steel from 16 to 12.2% and from 15.7 to 12.2% for 38XA steel.

The relative contraction of isothermomechanically treated rolled 35X steel with compression deformation from 4.9 to 17.7% decreases from 54 to 52%, and from 53.1 to 50.4% for 38XA steel.

According to State Standard 52643-2006, high-strength fasteners include products,whose tensile strength is greater than or equal to 800 MPa. Therefore, bolts and studs made of 35X and 38XA steels with the help of isothermomechanical method agree the strength class of high-strength fasteners. The results are presented in table 3.

Table 3 - Stud test results

Diameter,mm	Steel grade	Quantity, items.	σ_B, MPa	Ψ, %	δ, %	HB
M8	38XA	25	925	44,5	10,7	286
M8	35X	25	820	44,5	10,7	254

Hence, the test results analysis testifies that 35X steel can be used for bolts and studs with a strength class of 8.8. 38XA steel can be recommended for manufacturing hardware such as bolts and studs with the strength class 9.8g via cold die stamping after isothermal treatment of rolled stock.

Conclusion

Isothermomechanical treatment of rolled stock from structural steel of pearlite class 35X and 38XA with the diameter of 7.8 mm has been considered as a preparatiom stage for stamping studs and low-head bolts of strength class 8.8 and higher with hardening and tempering excluded from the process.

It has been proved that passing through thread rolling dies and surface hardening imparts such mechanical properties to M8 studs and bolts that they meet the requirements of strength classes 8.8 and 9.8 in compliance with State Standard 52643-2006.

Bibliography

[1] Pachurin G.V. Kinetics of fatigue failure of copper M1 and brass L63 // Non-ferrous metallurgy. Proceedings of the universities of the USSR. 1989. No. 1. - S. 96-101.
[2] Pachurin G.V. Fatigue failure at normal temperature of pre-deformed alloys. // Metallurgy and heat treatment of metals. 1990. No. 10. - S. 35-38.
[3] Pachurin G.V., Gushchin A.N. Improving the operational durability of metal products by technological methods // Herald of mechanical engineering. - 2007. - No. 6. - S. 62-65.
[4] Pachurin G.V., Gushchin A.N. Improving the operational durability of metal products by technological methods // Herald of mechanical engineering. - 2007. - No. 6. - S. 62-65.
[5] Pachurin G.V., Filippov A.A. Resource-saving and environmentally friendly metal surface treatment before cold heading // Ecology of Industry of Russia, August 2008. - P. 13-15.
[6] Pachurin G.V. The effect of temperature on the mechanical properties of sheet structural steels // Fundamental Research. - 2014. - No. 1. - S. 18-23.
[7] Pachurin G.V., Vlasov V.A. The mechanical properties of sheet structural steels at operating temperatures // Metallurgy and heat treatment of metals. - 2014. - No. 4 (706). - S. 48-53.
[8] Pachurin V.G., Galkin V.V., Pachurin G.V. Assessment of deformation heterogeneity in rolled products with a wedge profile // Fundamental Research. - 2014. - No. 11-4. - S. 765-773.
[9] Pachurin G.V., Filippov A.A., Pachurin V.G. Influence of temperature of isothermal processing with drawing on mechanical properties of hot-rolled steel 40X // Modern high technology. - 2015. - No. 2. - S. 98-106.
[10] Pachurin G.V., Filippov A.A., Pachurin V.G. The quality of the surface and structural condition of rolled products for hardware products made of steel 40X // Successes in modern science. - 2015. - No. 1-3. - S. 476-481.
[11] Rudskoy A.I. Lug: textbook. The allowance / A.I. Rudskoy, V.A. Lunev, O.P. Shaboldo - St. Petersburg: Publishing House Polytechnic. Univ., 2011 - 126 p.
[12] Filippov A.A., Pachurin G.V., Naumov V.I., Kuzmin N.A. The influence of the surface and structural state on the quality of rolled products for bolts // Fundamental research. - 2015. - No. 10-1. - S. 77-82.
[13] Filippov A.A., Pachurin G.V., Matveev Yu.I., Kuzmin A.N. Comparison of technological methods for preparing the structural and mechanical properties of a rolled surface for upsetting hardware in order to reduce the exposure of workers to hazardous and harmful factors // Basic Research. - 2016. - No. 10-1. - S. 88-96.
[14] Filippov A.A., Pachurin G.V., Kuzmin N.A., Matveev Yu.I. The method of forming the structural and mechanical properties of rolled steel for the planting of rod products // Ferrous metals. - 2018. - No. 4. - S. 36-40.
[15] Filippov A.A., Pachurin G.V., Kuzmin N.A., Matveev Yu.I., Deev V.B. Quality assessment of rolled products for cold forging // News of higher educational institutions. Ferrous metallurgy. - 2018. - Volume 61. - No. 7. - S. 551-556.
[16] Goncharova D.A., Pachurin G.V., Filippov A.A., Kuzmin N.A. Fatigue destruction of automo-bile construction materials // 3RD SCIENCE AND TECHNOLOGY SEMINAR "MOBILITY OF TRANSPORT AND TECHNOLOGICAL MACHINES", MTTM 2018, Nizhny Novgorod, August 28, 2018 - P. 012025.
[17] Guslyakova G.P., Zhbannikov S.I., Pachurin G.V. Fatigue failure resistance of deformed structural steels // Materials Science. - 1993. - T. 28. - No. 2. - S. 182-185.
[18] Pachurin G.V. Ruggedness of structural material and working life of metal components // Steel in Translation. - 2008. - T. 38. - No. 3. - S. 217-220.
[19] Pachurin G.V., Filippov A.A. Rational reduction of hot-rolled 40X steel before cold upsetting // Steel in Translation. - 2008. - T. 38. - No. 7. - S. 522-524.
[20] Pachurin G.V. Life of Plastically Deformed Corrosion-Resistant Steel // Russian Engineering Research. - 2012. - Vol. 32. - No. 9-10. - S. 661-664.
[21] Pachurin G.V., Vlasov V.A. // Mechanical properties of sheet structural steels at operating temperatures // Metal Science and Heat Treatment. - 2014. - T. 56. - No. 3-4. - S. 219-223.

IOP Conf. Series: Materials Science and Engineering **695** (2019) 012030 doi:10.1088/1757-899X/695/1/012030

[22] Pachurin G.V., Shevchenko S.M., Mukhina M.V., Kutepova L.I., Smirnova J.V. The Factor of Structure and Mechanical Properties in the Production of Critical Fixing Hardware 38XA // Tribology in Industry. 2016. - Vol. 38, - No. 3 - S. 385-391.
[23] Filippov A.A., Pachurin G.V., Naumov V.I., Kuzmin N.A. Low-Cost Treatment of Rolled Products Used to Make Long High-Strength Bolts // Metallurgist. - 2016. - Vol. 59. - Nos. 9-10. January. - S. 810-815.

Evaluation of medium factor influence on car materials fatigue resistance

D.A. Goncharova, G.V. Pachurin, A.A. Filippov, I.G. Trunova

Russia, Nizhny Novgorod, Nizhny Novgorod State Technical University named after R.E. Alekseev

E-mail: tugarino@mail.ru

Abstract. Parts and components of a car are subjected to various types of operational loads, causing the construction materials destruction. are vibration loads in corrosive environment are the most dangerous ones, they can increase the probability of road accidents.
The article considers the influence of corrosive medium on the patterns of fatigue resistance of automotive structural steels. The factors affecting the resistance to fatigue failure have been detected.
Corrosive medium makes it difficult to determine the moment of macrocrack onset and estimate the pre-onset period. However, using the curves of changes in the current deflection under sample cyclic loading, one can set down the moment of crack onset and determine its growth rate.
This allows us to justify the choice of materials for car parts and elements at the design stage, extend its life-span, reduce car and processing techniques material consumption, taking into account its service conditions.

Key words: structural materials, corrosive medium, fatigue cracks, deflection curves, cyclic durability.

Introduction

Parts and assemblies of a car are subjected to various types of operational stresses, that increases its damageability and leads to road accidents, which is especially dangerous in corrosive environment [9,10].

Structural material performance characteristics change at all stages of the metallurgical breaking range [15, 16, 19, 25]. High cost of metallic materials for car production gives a strong stimulus to increase reliability and life span of a car. Therefore, material fatigue characteristics studies [29] take priority in modern automotive industry, they help to optimize material choice at developing stage, to reduce metal consumption and to determine advanced technological processing methods [5, 6, 16, 24].

This involves studying the kinetics of metals and alloys fatigue fracture in corrosive environment, including determination of the crack onset protensity and the rate of their further growth, which allows to choose the material meeting operational requirements and, therefore, to prevent its destruction, and to reduce maintainance costs. However, the study of the process of material fatigue destruction is complicated by the duration and conditions of tests. At the same time, in the process of developing the part manufacturing technology it is important to take into account factors that affect its durability.

The purpose of the work carried out in NSTU named after R.E. Alekseev is to assess the influence of environmental factors on the fatigue resistance of automotive materials,that will help justify the choice of materials, increase the resource, reduce material consumption and develop manufacturing technology for producing car parts and elements ,which will consider operating conditions.

The object of research is automotive structural materials.

The subject of research is revealing of automotive structural materials behavior in the fatigue failure process, taking into account factors of cyclic loading conditions.

The influence of environmental factors in the tests

When designing automobiles, metal parts and elements from various structural materials obtained by various types and modes of processing are currently used. An important factor affecting their operational properties, in addition to the nature of the material,its structure and the quality of its surface, is the factor of the operating environment. Stress concentrators on the surface of the material contribute to the fatigue cracks onset and development. Under cyclic loads all processes in the surface layers of metal associated with the fatigue cracks onset are advanced, and all structural damage processes are activated at the surface.

In the automotive industry various alloys and structural steels are widely used. However, information on the influence of types and modes of their technological processing on their mechanical properties in corrosive media insufficient, which highlights the necessity of optimal design solutions and accurate use of technological processes in order to ensure the performance characteristics of parts and structures.

Therefore, the task of increasing the fatigue reliability of automotive parts with a possible decrease in metal consumption is urgent. Its solution involves the study of fatigue damage kinetics of automotive materials in corrosive environment, determination of the pre-crack period and the rate crack growth.

The solution to this problem will allow detecting the material that objectively meets operational requirements, which, in turn, will reduce the probability of structural damage under cyclic loads in corrosive environment and reduce repair costs.

As a rule, corrosion significantly accelerates fatigue processes in materials [11, 26], leading to a continuous decrease in durability in the multi-cycle region of fatigue curves that do not reach the breaking point.

Tests of smooth samples from steel 15KHN5DMF according to the scheme of zero clear bending at 167; 16.7 and 1.67 MHz frequencies showed [3] that sea water to a lesser extent affects the pre-crack period, but reduces the period of crack development. Comparison of the growth process of short and long cracks on steels A53CL1 [27] and St. JIS SNCM439 [21] revealed an acceleration short cracks growth up to 10 times against long ones.

With decrease in loading frequency, decrease in the durability in the medium is observed for the following materials: 15KHN5DMF [3], 422 SVS410 [20], low-alloy Cr-Mo steel [22], high-strength AISI 4340, 12Ni-5Cr-3Mo, 10Ni-Cr-Mo Co and martensitic steel with 18% Ni [23], 200 [26], 33NiCr [11], Nimonik-105 and IN738C alloys [28]. However, there is evidence that the effect of the loading frequency on the resistance to steel corrosion-fatigue fracture has non-monotonic character, for example, En56C and A533B-1 [17], austenitic steels 316L and 316S16 [31]. Sometimes there is a limit frequency [2] below which the development rate of a corrosion-fatigue crack in 12Kh2N steels either increases slightly or decreases.

The shape of the loading cycle does not have significant effect on the structural materials fatigue resistance in working environments at low voltages [11, 14], but its effect is enhanced at high voltages. The durability of the samples with a sinusoidal form of loading is less than with rectangular or trapezoidal forms. Thus, studies of 15KHN5DMF steel samples during a sinusoidal cycle in sea water by clear zero bending revealed [3] that a fatigue crack onset and growth proceed 10–90% more actively than during a trapezoidal cycle. This effect increases with the degree of cyclic deformation.

With an increase in the asymmetry of the loading cycle the resistance to material corrosion fatigue decreases the greater, the lower the stress intensity at the crack tip [11] is, the threshold stress intensity coefficient [18] and the difference between the crack growth rate in sea water and air [30] are decreased. However, for ship-hull steel 10KHSND for clean bending of frequent 35 Hz in natural sea water, increase in the asymmetry coefficient of the loading cycle does not affect crack onset and growth, but leads to decrease in their number.

An important factor is the degree of medium corrosiveness [11]. So, hydrogenation of metals, especially in high-strength steels with increased sensitivity to hydrogen embrittlement, is often decisive for alternating loading. The study of 08Kh17T and 12Kh18N10T stainless steels at a 2.5 Hz zeroing cycle showed [1] that in hydrogenated medium, the crack growth rate is a magnitude higher than in air.

The study of heat-resistant alloy Inconel-718 alloyed with molybdenum steel X20CrMo13 and titanium BT1-0 revealed [4] that one of the ways to increase metal products service life during fatigue in the medium is to use materials with high threshold stress intensity coeffici or technologies that increase it.

The simultaneous effect of alternating load and medium slows down the corrosion fatigue development due to the destruction of the passive film [13]. Due to the adsorption or adsorption-electrochemical effects of the medium, the initial phase of corrosion fatigue, as in corrosion cracking, is determined by local damage to the passive layer of the material surface. Direct microscopic observation of the working surface of the sample dangerous section [9] proved that corrosion-resistant failure begins in point damages, which occur mainly in non-metallic inclusions on resistant slide bands, the metal of which is most activated and is an anode for nearby surfaces.Then, these damages deepen, the metal thermodynamic activity increases due to the localization of stresses and the intensity of electrochemical processes. Clogging of pittings is obstructed by secondary corrosion products, slowing down oxidation under alternating loading.

It is a sourse of many corrosion-fatigue microcracks appeared on the surface of the samples without previously induced cracks, they locate close to each and exert cumulative effect on the distribution of the mechanical stresses fields. In this case, the cracks are not straight, but significantly branched.

Thus, according to reference data it difficult to determine the moment of macrocrack onset and estimate time interval before it.

Crack development [8, 14] is described by the "kinetic diagram of fatigue failure". It represents dependence curve of the "crack growth rate" *da/dN* to the "magnitude of the stress intensity coefficient" (SIC). The 2nd section of the KDFF is described by the Paris – Erdogan equation [8]:

$$da/dN = C \cdot \Delta K^{n}$$

where a is the length of the crack, m; N is the number of loading cycles; C is the coefficient; n is the exponent in the equation; $\Delta K = K_{max} - K_{min}$ is the SIC range, MPa$\sqrt{M}$.

The complete KDFF is complicated, therefore, its numerous descriptions are observed in the form of analytical dependencies that take into account the asymmetry of the cycle and describe the crack growth rate. The peculiarities of the KDFF configuration of many metals and alloys in corrosive media are studied in detail by A.N. Romanov [12].

However, the advantage of such kinetic approach is obvious only when macrocracks already exist in metal, the plane of their development is perpendicular to the direction of normal stresses, and the stress intensity at their peak can be controlled. The surface of smooth samples without previously induced cracks have many corrosion-fatigue microcracks, they locate close to each and exert cumulative effect on the distribution of the mechanical stresses fields. Therefore, it is almost impossible to assess the intensity of stresses at the crack peaks using existing methodological approaches. In addition, often the cracks are curved and significantly branched, which makes it extremely difficult to apply the basic hypotheses of fracture mechanics to the analytical description of fatigue kinetics in corrosive medium.

In the course of experiment using the optical observation scheme for the cylindrical sample fatigue surface (Fig. 1), it was ascertained [8, 26] that the deflection curves represent an integral characteristic of destructive changes under alternating material loading.

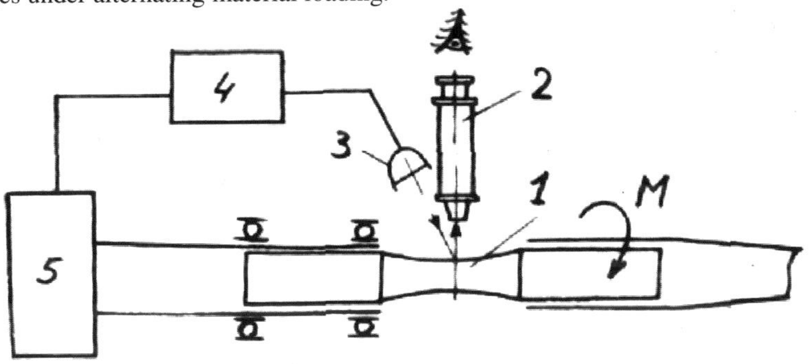

Figure 1. The scheme of studying the sample fatigue surface: 1 - sample; 2 - a microscope; 3 - stroboscopic lamp; 4 - strobe; 5 - phase synchronizer

It was shown [5, 24] that hardening is immediately observed in materials after annealing, accompanied by intense sliding in almost all grains, as evidenced by deflection decrease (Fig. 2) [5]. Then deflection stabilizes, which is associated with the developed slip along the secondary bands, the intersection of the bands, and ends with a macrocrack that grows untill the sample ruptures.

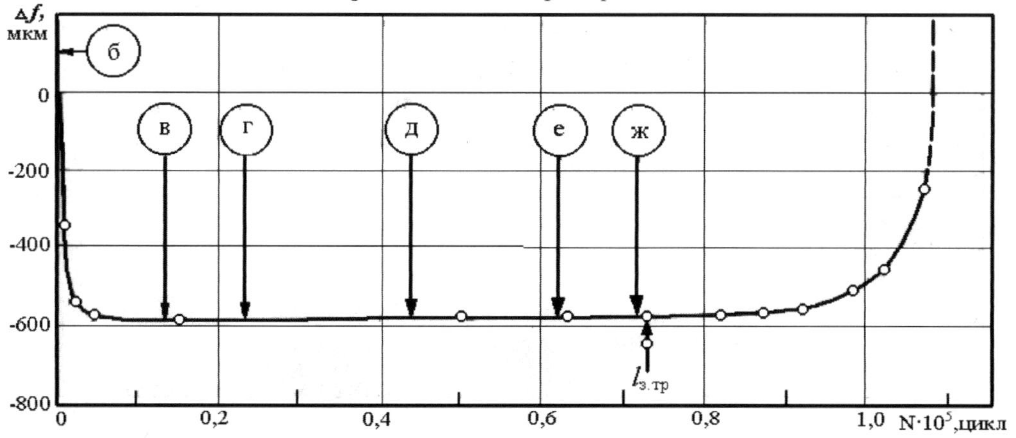

Figure 2. Change in the current deflection at different stages of cyclic loading and the moment of a macrocrack onset in annealed copper sample C1 for σ_a = 140 MPa (% of *N*p): *a* - initial state; *б* – 0,63%; *в* - 12,5%; *г* –22,0%; *д* – 40,7%; *е* –57,6%; *ж* –66,8%

Technologically hardened materials behave differently, their deflection increases immediately (Fig. 3), which corresponds to the appearance of rare slip bands in some grains.

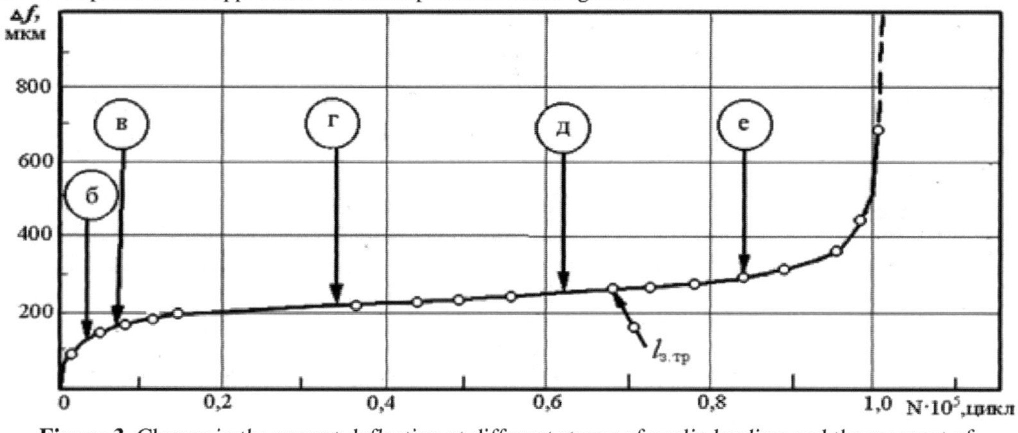

Figure 3. Change in the current deflection at different stages of cyclic loading and the moment of macrocrack onset in solid brass sample L63 at σ_a = 300 MPa (% of Np): *a* - initial state; *б* –3,4%; *в* – 6,7%; *г* – 33,5%; *д* – 61,3%; *е* –83,8%

Further, during the deflection stabilization phase, compacted sliding along the primary and secondary planes is observed. The phase of increased deflection growth is characterised by the surface crack onset with the length of about 1.0 mm [6].

The outcomes and their consideration.

The results of mathematical processing of a typical deflection curve in corrosive environment (3% aqueous solution of sea salt) under symmetrical cantilevered bending with the rotation of a model cylindrical sample from a technologically hardened B63 at σa = 250 Mpa are appended in the table.

Table 1

The results of mathematical processing of a typical deflection curve in a corrosive environment with symmetrical cantilevered bending with the rotation of a model cylindrical sample from technologically hardened brass B63 at σa = 250 MPa.

σ_a, MPa	$\varepsilon_{\text{пр}}$, %	N_p, cycle	$N_{з.\,тр.}$, cycle	$N_p - N_{з.тр}$, cycle	l_s, mm	$V_{тр.}$, mkm/ cycle
B63, cold-rolled, air						
300	0	341000	270000	71000	3,66	$5{,}15 \cdot 10^{-2}$
300	25	115000	80000	35000	3,34	$9{,}54 \cdot 10^{-2}$
B6, cold-rolled, corrosive medium						
300	0	51000	40000	11000	3,10	$48{,}2 \cdot 10^{-2}$
300	25	68100	54000	14100	3,31	$23{,}5 \cdot 10^{-2}$

Note:
σ_a –stress, MPa;
$\varepsilon_{\text{пр}}$ – tensile strength ZD 10/90;
deformation rate *$5{,}6 \cdot 10^{-3}\, c^{-1}$, %*;
N_p – full durability, c;
$N_{з.\,тр.}$ – number of cycles before crack onset, cycle;
$N_p - N_{з.тр}$ – number of "oparational" cycles with a crack, cycle;
l_s – the length of the sample's pure fatigue fracture, mm;
$V_{тр.}$ – crack growth rate, mkm/cycle.

Full durability *Np* and durability before crack onset *N3.mp.*when cold-rolled brass is tested in air, (3.41.105 and 2.7.105 cycles, respectively) is higher than that of cold-rolled brass with a deformation of 25% (1.15.105 and 8.104 cycles, respectively). However, the crack growth rate in cold-rolled B63 ($5{,}15 \cdot 10^{-2}$ mkm / cycle) is vice versa lower than that in cold-rolled and deformed by 25% ($9{,}54 \cdot 10^{-2}$ mkm / cycle).

In the medium both full cyclic durability and durability before macrocrack onset is lower, and its growth rate is higher than in air, which confirms the conclusions [7–9, 11].

When tested in a medium, the full durability and the durability before crack onset of cold-rolled brass with additional deformation of 25% ($6{,}81 \cdot 10^4$ and $5{,}4 \cdot 10^4$ cycles, respectively) higher than that of just cold-rolled one($5{,}1 \cdot 10^4$ and $4{,}0 \cdot 10^4$ cycles, respectively). On the contrary, the crack growth rate is higher for cold-rolled B63 ($48{,}2 \cdot 10^{-2}$ mkm / cycle) compared to cold-rolled brass with additional deformation of 25% ($23{,}5 \cdot 10^{-2}$ mkm / cycle).

Conclusion

Medium influence, its aggressiveness, frequency, form and asymmetry of the loading cycle on the resistance to fatigue failure has been revealed and estimated.

It has been established that materials and types of automobile parts processing, whose pre-onset crack resource constitutes the larger part of their "life" till the final destruction, are preferable, that allows to detect a crack in the course of regular technical inspection and improve the maintainability of a car.

It has been proved that the sample deflection curves make it possible to track the moment of fatigue crack onset and determine its growth rate, which is especially important in corrosive medium, since direct observation of the sample surface is impossible.

Using the data obtained, it is possible even at the design stage to make the optimal choice of material and manufacturing technology, and at the maintenance stage to prevent the destruction of structural elements and parts that are subjected to operational cyclic loads.

Bibliography

[1] Vitvitsky V.I., Levitsky M.O., Duryashin et al. // Phys. – Chem. fur. materials. - 1981. - 17, - No. 3. - S. 107-108.
[2] Magdenko A..N., Sadichenko V.I. The influence of the cycle shape on the growth rate of a corrosion crack in 12Kh2N steel at low-cycle fatigue. // Phys.-Chem. fur. materials. - 1984. - No. 5. - S.102-103.
[3] Maksimovich G.G., Kobzaruk A.V. The origin and development of low-cycle fatigue cracks in steel 15KHN5DMF in sea water. // Phys. - chem. fur. materials. - 1984. - No. 5. - S. 16-20.
[4] Nironovich I.A. // Phys.-Chem. fur. materials. - 1980. - 18, 5. - No. 5. - S. 22-25.
[5] Pachurin G.V., Gushchin A.N., Galkin V.V., Pachurin V.G. The theoretical basis for increasing the operational durability of stamped metal products: textbook. A manual for university students / NSTU. N. Novgorod, 2006. - 173 p.
[6] Pachurin G.V. Fatigue resistance at different temperatures of annealed and hardened brass L63 // International Journal of Experimental Education. - 2014. - No. 3-1. - S. 119-126.
[7] Pachurin G.V. The role of surface structure in corrosion fatigue of deformed metal materials // Modern problems of science and education. 2014. No. 1; URL: http://science-education.ru/ru/article/view?id=11907 (accessed: 08/10/2018).
[8] Pachurin G.V. Corrosion Durability of Deformation-Hardened Metal and Alloy Products: A Training Manual. - 2nd ed., Ext. - St. Petersburg: Publishing House "Lan", 2014. - 160 p.
[9] Pachurin G.V., Kudryavtsev S.M., Soloviev D.V., Naumov V.I. The body of a modern car: materials, design and production: Textbook / Ed. G.V. Pachurina. - 3rd ed., Revised. and add. - St. Petersburg: Publishing House "Lan", 2016. - 316 p.
[10] Pachurin G.V., Galkin V.V., Pachurin V.G. Durability of hardened metals and alloys: a training manual / G.V. Pachurin, V.V. Galkin, V.G. Pachurin. Stary Oskol: TNT, 2017. - 228 p.
[11] Pohmursky V.I. Corrosive fatigue of metals. - M.: Metallurgy, 1985. - 207 p.
[12] Romanov A.N. // Phys.-Chem. fur. materials. - 1980. - 16. - No. 2. - S. 27-30.
[13] Sinyavsky Sun., Valkov V.D., Budov V.M. Corrosion and protection of aluminum alloys. - M.: Metallurgy, 1986. - 368 p.
[14] Terentyev V.F., Petukhov A.N. Fatigue of high strength metal materials. - M.: IMET RAS - TsIAM, 2013.515 s.
[15] Filippov A.A., Pachurin G.V., Kuzmin N.A. et al. Quality assessment of rolled steel for cold forging // News of higher educational institutions. Ferrous metallurgy. - 2018. - Volume 61. - No. 7. - S. 551-556.
[16] Filippov A.A., Pachurin G.V., Kuzmin N.A. The method of forming the structural and mechanical properties of rolled steel for the planting of rod products // Ferrous metals. - 2018. - No. 4. - S. 36-40.
[17] Atkinson L.D., Lindley T.S. The influence of the frequency [loading] and temperature on the growth of a fatigue crack under the influence of a medium in steels [at stresses] below KISCC. // Influence Environment Fatique Conf., London, 1977. - S. 65-74.
[18] Chen Daolun, Wang Zhongguang, Jiang Xioaxia, Ai Suhua, Shi Changxu. Fatigue crack growth in rolled two-phase steel // Steel Res. - 1989. - 59. - No. 7. - S. 319-322.
[19] Filippov A.A., Pachurin G.V., Naumov V.I., Kuzmin N.A. Low-Cost Treatment of Rolled Products Used to Make Long High-Strength Bolts // Metallurgist. 2016. Vol. 59. Nos. 9-10. January. S. 810-815.
[20] Komai Kenjiro, Kanasari Hiroshi. Corrosion fatique crack growth of a martensitic stainless steel in NaCL solution. // Bull. JSME. - 1985. - 28. - No. 236. - S. 202-208.
[21] Nakai Yoshikazu, Akagi Hedenari, Kitamura Yasuhiro, Ohji Kiyotsugu. The growth of short surface cracks in high-strength low-carbon steel under cyclic loading in a 3.5% NaCL solution. // Nihon Kikai Takkai rhombunsu. A = Tgank.Jar.Soc.Mesh. Eng. A. - 1989. - 55. - No. 516. - S. 1724-1731.
[22] Nishyima Satashi, Abe Takayuki, Mashuda Chitoshi, Hirukawa Hisashi. Characteristics of corrosion fatigue and the influence of the frequency on it for low alloy steels in salt water during rotation by bending. // Nihon Kikai Takkai Rhombus Trans. Jap. Soc. Mesh Eng. 1985. A51. No. 461, S.156-160.
[23] Nagai Kin-ichi. Corrosion resistance and fracture mechanics. // Yessetsu Gakkashi, J.Jap. Well. Soc. 1976. 55. No. 11. S.926-937.
[24] Pachurin G.V., Goncharova D.A., Filippov A.A. and etc. Development of fatigue test technology of sheet automo-bile materials // Eastern-european jour-nal of enterprise tech-nologies ISSN 1729-3774. - Vol 5, No 12 (95). - 2018. - S. 31-37.

[25] Pachurin G.V., Shevchenko S.M., Filippov A.A. and etc. Defining Role Metal Performance for cold bolt upsetting (bolt head) // International Journal of Mechanical Engineering and Technology. - 2018. - No. 9 (4).
[26] Pachurin G.V. Life of Plastically Deformed Corrosion-Resistant Steel // Russian Engineering Research. - 2012. - Vol. 32. - No. 9-10. - S. 661-664.
[27] Pei Hongxun, Yang Jigjun, Ke Wei. The growth of short cracks in A537CLl steel under cyclic stress in a 3.5% NaCL solution. // Jin-shu-snabao = Acta met. sin -1988. - 24. - No. 6. - S. B393-B397.
[28] Pettit D.E., Ryder J.T., Krupp W.E., Hoeppner Scarlin R.V. Influence of loading frequency and medium on crack growth during high-temperature fatigue of nickel-based alloys. // Adv. Res. Strength Fract. Mater. 4th Int. Conf. Fract., Waterloo, 1977. New York E.A. - 978. - S. 349-857.
[29] Schmidtmann Eugen, Wirths Dieter Aachen. Influence of loading frequency on crack propagation characteristics in high-strength 33NiCrMo145 steel under conditions of pulsating tension in various media. // Arch Fisenhuttenw. - 1973. - 49. - No. 10. - S. 483-437.
[30] Sumita Masae, Maruyama Norio. The propagation of fatigue cracks in the Ti-6A1-4V alloy in sea water at low LC values. // Tetsu to Hagane, J. Iron and Steel Inst. Jap. - 1989. - 74. - No. 9. - S. 1854-1861.
[31] Smethurst E., Waterhouse R.B. Effect of frequency on the fretting fatique behavior of two austenitic stainless steel implant materials in Hanks solution. // Proc. 2th Int. Conf. Mesh Behav. Mater, Boston, Mass., 1976. - S.1, 1976. - S. 695-699.

The consideration of road longitudinal slopes in automobile fuel consumption rationing

G.V. Borisov[1], N.A. Kuzmin[1], J.I. Molev[1], I.R. Mavleev[2] and I.I. Salakhov[2]

[1] Nizhny Novgorod State Technical University named after Alekseev R.E., Nizhny Novgorod
[2] Kazan Federal University

E-mail: kafedraat@gmail.com

Abstract. Nowadays the consideration of longitudinal slope when rationing automobile fuel consumption is hardly taken into account, and it is received just impliedly when paying attention to the area relief. Along with this, the influence of slopes of flat and slightly hilly terrain on fuel consumption is not taken into consideration at all. This article presents an easy calculation approach to the consideration of the quantitative effects of road slope angels (rise and descent) on the final fuel consumption of a vehicle on real routes. The notion of a reduced slope angle is inroduced,a substantial increase of fuel consumption on descents in comparison with rises is ascertained.

Keywords:*fuel consumption, fuel consumption rationing, the problems of rationing, road slope.*

In Russia longitudinal slopes of roads (rises and descents) are taken into account in the rationing system of automobile fuel consumption impliedly, by means of adequate allowances when calculating the operational rate of fuel consumption [1,2].

Thus, in July 1, 1960 in the USSR recommendatory regulatory document "Standard quotas of liquid fuel consumption for cars" became effective ,it was approved by the USSR Council of Ministers Resolution on March 5, 1960 No. 252 "On quotas of liquid fuel consumption for cars", according to which:

- when working on roads in mountain areas (more than 1,500 meters above sea level) or on roads with a difficult plan (having an average 1 km track with more than five curves with the radius less than 40 meters): in the summer time - up to 10 percent and in the winter time - up to 20 percent;

Besides turnings, the term "complex road scheme" in the document implied longitudinal slopes, which can be found on the vast majority of mountain roads. However, no recommendations for specific allowances taking slope rate into consideration, when fuel rationing, were meant. It should be pointed out, that such recommendations didn't appear in all the subsequent regulatory documentation on the automotive fuel rationing.

Later on, the Decree of the Ministry of Transport of the Russian Federation dated March 14, 2008 No. AM-23-p "Automotive fuel and lubricants consumption quotas" fixed the following allowances, which also took longitudinal slopes on the routes into account only impliedly :

- operation of vehicles on public roads (categories I, II and III) in the mountains, including cities, towns and suburban areas, at an altitude above sea level:

• from 300 to 800 m - up to 5% (lower mountains);
• from 801 to 2000 m - up to 10% (mid mountains);
• from 2001 to 3000 m - up to 15% (high mountains);
• over 3000 m - up to 20% (high mountains);

- the operation of vehicles on public roads of categories I, II and III with a complex area relief (outside cities and suburban areas), where an average 1 km track has more than five curves with the radius less than 40 meters (or per 100 km ways - about 500) - up to 10%, on public roads of categories 1V and V - up to 30%.

This regulatory document doesn't take proper account of longitudinal slopes on roads in flat and slightly rolling country (reliefs P1 and P2 with heights above sea level up to 200 m and from 200 to 300 m respectively) in fuel consumption rationing.

The latest regulatory adjustment in the sphere of automotive fuel and lubricants consumption rationing was made on May 14, 2014 by the Decree of the Ministry of Transport of the Russian Federation No. NA-50-r "Alteration of recommended practice "Automotive fuel and lubricants consumption quotas "attached by the Ministry of Transport of the Russian Federation Decree on March 14, 2008 No. AM-23-p", and then it became operative with the Ministry of Transport of the Russian Federation Decree on July 14, 2015 N HA-80-p, according to which new models with corresponding basic standards were added to the list of vehicles. No other changes were made, including adjustments for considering longitudinal slopes of the roads in automotive fuels rationing.

Factors affecting fuel consumption make up the so-called fuel balance of a car [1,3]. It includes the following components : mechanical, thermal and pumping losses in the engine ($Q_{ДВ}$); mechanical losses in transmission (Q_{TP}), as well as everything regarding road resistance. Road resistance is composed of rolling resistance (Q_f), aerodynamic resistance, i.e. air resistance (Q_W) and inertia overcoming (Q_γ). The listed factors affecting fuel consumption are also taken into account in rationing methods, determining the basic (linear) fuel consumption rate, or are taken into account by the corresponding fixed allowances. Among other components in the fuel balance, there is one that determines losses on road longitudinal slopes overcoming (Q_α). These losses are not taken into account in regulatory documents on automotive fuel consumption rationing. Thus, even fuel balance analysis suggests that the standards for automotive fuel consumption rationing need revision and completing.

In this regard, the Department of "Automobile Transport" NSTU named after R.E. Alekseev has been carrying out research on of longitudinal slopes effect on automotive fuels consumption, aimed at revealing quantitative relationships between these factors.

Such interdependence is recognized by some scientists of the sphere, but despite a wide range of scientific works, textbooks, teaching aids and other technical documentation, this fact is not reflected in the system of automotive fuel and lubricant consumption rationing.

The fuel consumption when a vehicle is moving on slope is increasing due to gravity $F_i = G_a \sin\alpha$, called lifting resistance [4], counter- or co-directional (respectively, positive grade - rise and negative grade - descent) to movement (Fig. 1) .

To study the quantitative relationships between the slope angle and fuel consumption, a GAZelle Next car of class N1 (up to 3.5 tons) model A21R22 with a Cummins 2.8 ISF engine and a 5-speed manual gearbox GAZ was taken.

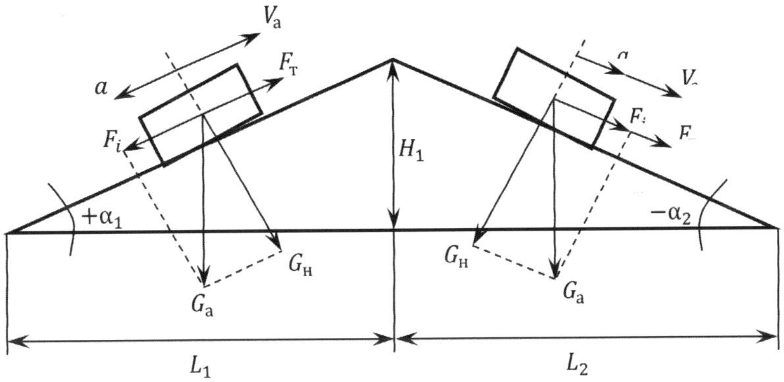

Fig. 1. The model of car motion on the slope:
a – acceleration; V_a – speed; F_i – lift resistance;

$F_т$ – tractive force; G_a – weight; $G_н$ – normal weight;
H_1 – height of lift; L_1,L_2 и $+\alpha_1$, $-\alpha_2$– lengths and angles of rise and descent.

To reveal this relationship the fuel consumption was calculated on the horizontal section L_1 + L_2 and the corresponding sections of rise and descent. According to construction regulations 2.05.02-85, the maximum road slope grade (i = tgα = H / L) is 0.03-0.07, particularly complex relief areas, slope rise 0.02 is allowed. The maximum admissible (in road construction) slope is 0.07-0.09, which corresponds to 4,2-5,5° (the value α = 5,5° is taken in this work). The movement in both cases will be characterized by different values of the power P_ψ needed to overcome total road resistance.

Obviously, there are several options for the expected result (output):

1. The power expended in overcoming rise and descent resistance is equal, i.e. the increase in power losses on the rise is compensated by their equivalent decrease on the descent. In this case, the total fuel consumption will be equal to the consumption when driving on a horizontal road section, and the relationship cannot be observed. This is exactly how the present fuel rationing system works for roads in flat and slightly hilly terrain, and therefore road slopes are not taken into account in regulatory documents on fuel rationing.
2. The powers expended in overcoming rise and descent resistance are not equal, and which is more, their sum is positive. In this case, the total consumption will be greater than that on horizontal road, and the “rise-descent” model comes to “rise” movement with a smaller angle (it can be designated as the reduced angle of the road).
3. Similarly, the powers are not equal, but their sum is negative. The final consumption will be less, and the desired model comes to “descent” movement with negative reduced slope angle.

As is known [3, 4], the total road resistance ψ takes into account the rolling resistance (in the form of the coefficient of rolling resistance f) of a wheel on supporting surface and the angle of the road longitudinal slope α through the expression ψ = f · cosα + sinα (for horizontal roads ψ = f) . In these research, asphalt covered roads are meant, for horizontal profiles with $\psi = f_{ac\phi}$ = 0,008– 0,015 [3].

Power expended in overcoming road resistance is determined by the expression $P_{\psi} = F_{\psi} \cdot V_a$, where $F_{\psi} = F_f + F_{\alpha} = G_a \cdot f \cdot \cos\alpha + G_a \cdot \sin\alpha$ is the resistance to movement. For horizontal roads, these power losses will amount to P ψ = 6.3 kW. In this article we use the value of the coefficient f obtained in the course of research work carried out by students of the MA course at the Department of Automobile Transport, Nizhny Novgorod State Technical University. R.E. Alekseev, it was calculated for the road conditions of N.Novgorod and forms $f_{ac\phi}$ = 0,011. This value was obtained as a result of several measurements with the help of a car weeling-out method, using the expression $f = \frac{|a|}{g}$, where a is the deceleration when weeling out, modulo [1,3].

Using the steady-motion fuel consumption equation at a speed of V_a = 60 km / h, the fuel consumption of a GAZelle Next car with a total mass of 3.5 tons will be Q_s = 9.33 l / 100 km (according to official data from the manufacturer of the GAZ car the fuel consumption for the specified car of laden mass at the appropriate speed will be 8.5 l / 100 km).

Further calculations showed that the average fuel consumption of the vehicle in the “rise-descent”sections is 18.35 l / 100 km, provided that the requirements for road safety are met, which appeared primarily in inadmissibility of inertial motion (without gear) when descending. Thus, part of the engine power released on the descent will be almost completely dissipated through the car braking mechanisms [5,6]. Mostly this determines a significant increase in fuel consumption for overcoming slopes in general compared with driving along horizontal sections of the road.

Hence, if rise and descent characteristics are equal ($\alpha_1 = \alpha_2$) ,the calculated values testify to the second variant of the above conclusions or expected results. This allows us to reduce the rise – descent model (Fig. 1) to the “rise” model and to determine the slope angle at which the fuel consumption will correspond to the fuel consumption in the “rise – descent” model. In this work, it is convenient to designate this calculated angle as $\alpha_п$ - reduced slope angle (Fig. 2).

Calculations have proved that for the above conditions of the problem with $|\alpha_1| = |\alpha_2|$ = 5,5° the rise – descent model can be reduced to the “rise” model with a given angle $\alpha_п$ = 1,134°.And the total fuel consumption will be the same.

In the practice of fuel consumption rationing, when routes have many slopes with different characteristics (height and length) a general expression for calculating the reduced slope angle can be derived, it can be either positive or negative and, accordingly, will increase or decrease the final value of fuel consumption.

For the entire route having many slopes it is necessary to summarize their values ,somehow, by the values of the angles. In this case, a final angle, most likely, should be called accumulated reduced angle of the slopes. At present it will not be particularly difficul to put this idea into practice with the help of the software available that uses satellite navigation data (for example, the GoogleEarth program), which ensures that slope characteristics can be determined with any resolution.

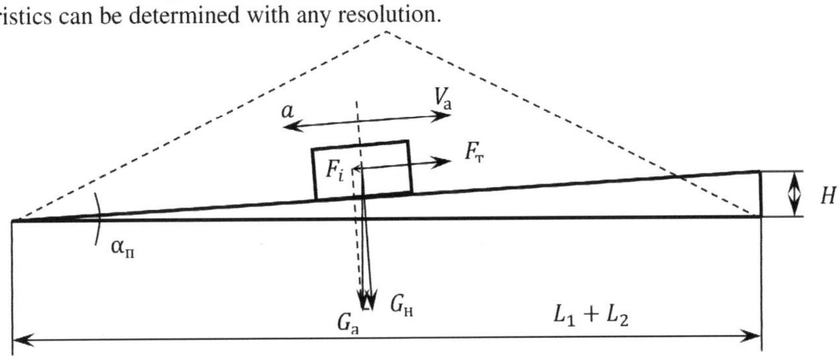

Figure 2. The slope angles calculation model for fuel consumption rationing:
$\alpha_п$ – reduced slope angle.

The output of the research described is a mathematical model of accumulated reduced slope angle for automotive fuel consumption rationing considering all the factors that influence fuel consumption, including losses for road slopes overcoming.

References

[1] **Kuzmin N. A.** *Tekhnicheskaya ekspluataciya avtomobilej: normirovanie i upravlenie: uchebnoe posobie*.[The technical operation of vehicles: regulation and management: tutorial]. N. A. Kuzmin.– M.: FORUM. - 2011. – 224 p.

[2] **Kuzmin N. A.** [The problems of rationing of expenses of automobile fuels and lubricant materials in the Russian Federation]. *Problemy normirovaniya raskhodov avtomobil'nyh topliv i smazochnyh materialov v RF//* Avtotransportnoe predpriyatie. – 2010. - №8. – p. 20-22

[3] **Kuzmin N. A.** *Teoriya ekspluatacionnyh svojstva avtomobilya: uchebnoe posobie*[The theory of operation properties of avehicle: tutorial]. N. A. Kuzmin, V.I. Peskov. – M. Forum, NIC INFRA-M. – 2013. – 256 p.

[4] **Kravec V. N.** *Teoriya avtomobilya. Uchebnik dlya vuzov*.[The car theory. Textbook for universities].V.N. Kravec, V.V. Selifonov. – M.: OOO «Grinlajt+». – 2011. – 884 p.

[5] **Borisov, G.V.Leliovskij K.Ya., Kuzmin N.A.** [On the issue of rationing the consumption of liquid fuels in road transport]. *K voprosu o normirovanii raskhoda zhidkih topliv na avtomobil'nom transporte*.G.V. Borisov, // Avtotransportnoe predpriyatie. – 2015. - № 2. – p. 51-55

[6] **Kuzmin N. A. Korchazhkin M.G.** [The features of the operation of city buses on routes with rises]*Osobennosti ekspluatacii gorodskih avtobusov na marshrutah s pod"emami //* Gruzovoe i passazhirskoe avtohozyajstvo. – 2005. – №12. – p. 28-31.

IOP Conf. Series: Materials Science and Engineering **695** (2019) 012033 doi:10.1088/1757-899X/695/1/012033

On Comparing Radar Performance and Conducting Their Assessment Tests

Myakinkov A. V., Fadeev R. S., Burov V. N., Yarjemsky A.D., Rogov P. S., Trusov Y. P., Toropov E.I., Moshkov P. S., Kolesnitcenko D.A.

Nizhny Novgorod State Technical University named after Alekseev R.E., Nizhny Novgorod

E-mail: petr.rogov@nntu.ru

Abstract. The present article deals with the initial stage of integration of the automatic emergency braking system to the vehicle. The analysis of directives and regulations demands is carried out. The methodology of comparative testing of automobile radars as a part of a vehicle is clearly shown. Basing on the test results, the authors tried to estimate the influence of environmental elements of an automotive radar installed in a car on basic radar characteristics for detecting targets. Further actions are suggested to evaluate the performance of car radars necessary for rational selection of the model itself and the manufacturer of this sensor, which would be important for subsequent development and integration of automatic braking systems and adaptive cruise control.

Key words: ADAS, Automatic Emergency Braking, automotive radar, object detecting, testing, comparison.

Many active driver assistance systems (ADAS), which control speed and direction of the vehicle in automatic mode, are becoming a typical option in a modern car. Taking into consideration positive influence on improving road safety, many ADAS systems are included in the list of vehicle systems required to obtain vehicle type approval and, after the transition period is over, their presence is necessary for vehicle registration. At present, in the countries of the European Union (EU), for registration of vehicles of categories N2, M2 (class B), N3, M3 (class III), it is required to have AEB (Automatic Emergency Braking) [1] and LDWS (Lane Departure Warning System - warning about an unintentional exit from a lane) [2]. The LDWS function does not imply control of the speed and direction of movement of the vehicle and only performs a driver warning. The AEB function must control the deceleration of the car and, accordingly, interact with the braking system, as well as interrupt the fuel supply to the internal combustion engine (ICE) for more effective braking. For cars with automatic transmission, it is also possible to interrupt the transmission of the torque to the wheels of the car by performing control signals from the AEB function. Thus, in the event of misfiring or malfunction resulting in incorrect control signals, the AEB function is potentially more dangerous than LDWS, due to its control interaction with the main vehicle systems. For this reason, all design decisions related to the integration of the AEB function in the vehicle should be carefully checked, including through functional and road tests.

Choosing an AEB detection sensor or a combination of sensors is the first fundamental step in implementing a system. The more justified is the choice of the type and model of a detection sensor at the stage of developing a system concept, the less likely are changes in the system and its environment during development, which in turn entails significant time and material costs. According to the data from open sources and basing on the functional features of various sensors [3-5], the main sensor for detecting the AEB function on foreign cars is a radar located in the elements of the front of the car [6-9]. For this reason, the present work is devoted to solving the problem of rational choice of the radar model, taking into account its functioning in the vehicle, as well as the minimum functional assessment of the elements of the environment of the radar in a car.

The main performance characteristics of automobile radars are [5]: a distance range and angle range (azimuth and elevation), distance (angle and velocity) measuring resolution, accuracy of measuring, velocity range. The current regulatory document that regulates the functional requirements for the function of the AEB for the Russian Federation is UN Regulation No. 131 [10]. Therefore, the provisions of the Rules were taken into account when developing a test procedure for checking the above characteristics of automobile radars.

Functional tests were carried out at a specialized training ground. Functional races for a vehicle equipped with a radar are shown in Fig.1. Before the start of the functional phase of the tests, the vehicle was accelerated to a speed of 80 km / h. The functional stage of the tests ended when the indicated distance was reduced to the minimum value at which the vehicle can be dodged or braked to avoid a collision with the target. To check the detection distance R and the viewing angle in azimuth φ, the target in different races should be located differently in the places marked with numbers in Figure 1. Verification of the vertical viewing angle is possible when changing the radar setting on the car - turning it 90 °. Functional rides are performed separately for stationary purposes (car, pedestrian) and separately for moving targets (car speed of 12 km / h, pedestrian - 5 km / h).

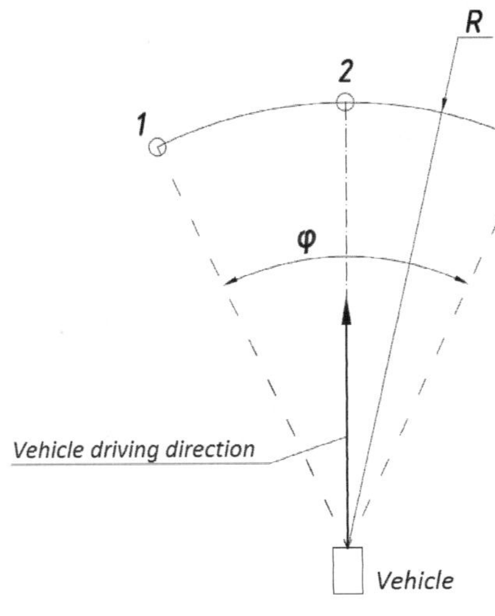

Figure 1. Scheme of functional races when testing automobile radars as part of a vehicle

Verification of the minimum detection distance by radar is performed by approaching the vehicle with the target (in accordance with Figure 1) at the lowest possible steady speed. After the target disappears from the radar's field of view, the functional vehicle slows down to a complete stop.

Two radars participated in the tests examined: Contienental ARS 408 and MARS –2A1 / 10 (manufactured by Milander). A CAN adapter was used to communicate with the Contienental ARS 408 radar. The MARS – 2A1 / 10 radar was connected to a laptop computer via an Ethernet connection. A multifunctional speed meter Racelogic VBOX 3i RTK was used to measure the dynamics of a functional vehicle. The change in the relative position between the functional vehicle and the target was recorded using two telemetry kits and a Racelogic DGNSS RTK base station. The listed equipment and radars were checked sequentially in the design of the GAZelle Next and GAZon Next automobiles. Fragments of testing and installing measuring equipment on prototypes are shown in Fig. 2-3.

a)

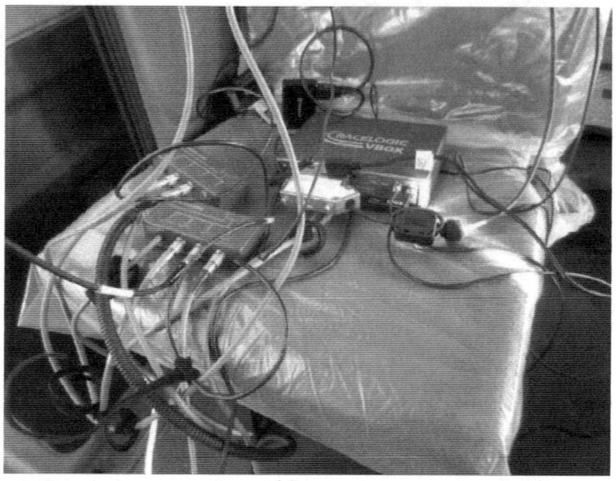
b)

c)

Figure 2. Fragments of functional tests of radars in GAZelle Next: a,b – Installation measuring equipment; c – General view of the prototype with an automotive radar installed in the front

According to the results of processing data from radars, after testing, estimates of the trajectories of the objects (car, pedestrian) in the Cartesian coordinate system with the center at the location of the radar were obtained. From the presented results of detection and trajectory tracking of objects, it can be seen that the car is easily detected by radars at distances of 250 m or more (Fig. 3), and the pedestrian - at distances of up to 100 m (Fig. 4).

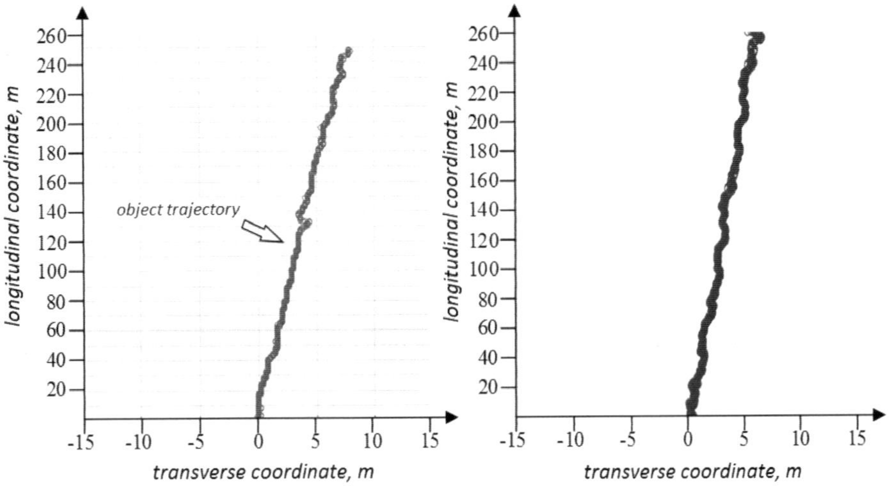

a) b)

Figure 3. The target of the detection is a car. Results of the trajectory tracking of the car with frontal radars : a) ARS 408, b) MARS –2A1/10

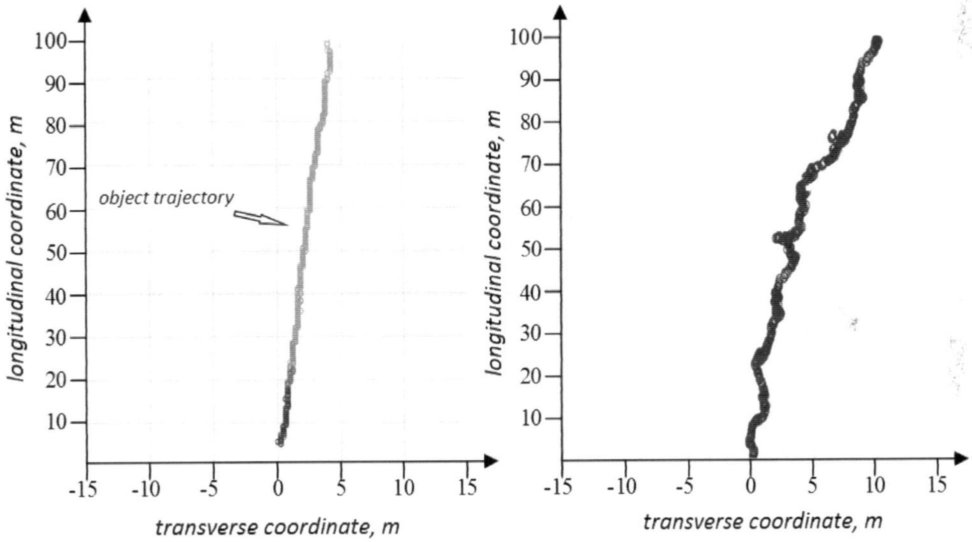

a) b)

Figure 4. Results of the trajectory tracking of the pedestrian with frontal radars: a) ARS 408, b) MARS –2A1/10

The results obtained from the tests led to the conclusion that the estimated characteristics of the radars are fully consistent with the information specified by the manufacturer. The environment elements of radars installed on cars do not reduce the performance of these components.

The next step in assessing the performance of car radars is to test them as part of a car on public roads. In this case, the car must be equipped with a video camera. Video recording must be synchronized in time with the transmission of data from the radar. This test setup allows you to estimate the number of false targets captured by the radar, as well as the detection of targets that present a danger of collision in more difficult road conditions, compared with the polygon tests.

Experimental studies were performed using the equipment of the Center for the collective use of NSTU "Transport Systems".

The studies were carried out with the financial support of the Ministry of Science and Higher Education of Russia in the framework of the project "Creation of a high-tech production of safe export-oriented GAZ cars with autonomous control elements and the possibility of integration with the electric platform on the

IOP Conf. Series: Materials Science and Engineering **695** (2019) 012033 doi:10.1088/1757-899X/695/1/012033

basic components of Russian production" under contract No. 03.G25.31.0270 of 05.29.2017. (Decree of the Government of the Russian Federation of April 9, 2010 No. 218).

References

[1] Commission Regulation (EU) No 347/2012 of 16 April 2012 implementing Regulation (EC) No 661/2009 of the European Parliament and of the Council with respect to type-approval requirements for certain categories of motor vehicles with regard to advanced emergency braking systems // EUR-lex. Access to European Union law URL: https://eur-lex.europa.eu/legal-content/EN/TXT/PDF/?uri=CELEX:32012R0347&from=en (application date: 25.08.2019).

[2] Commission Regulation (EU) No 351/2012 of 23 April 2012 implementing Regulation (EC) No 661/2009 of the European Parliament and of the Council as regards type-approval requirements for the installation of lane departure warning systems in motor vehicles // EUR-lex. Access to European Union law URL: https://eur-lex.europa.eu/eli/reg/2012/351/oj (application date: 25.08.2019).

[3] Characteristics of the stereo video camera // Bosch Mobility Solutions URL: https://www.bosch-mobility-solutions.com/en/products-and-services/passenger-cars-and-light-commercial-vehicles/driver-assistance-systems/lane-departure-warning/stereo-video-camera/ (application date: 22.08.2019).

[4] HDL-32E // Velodyne Lidar URL: https://velodynelidar.com/hdl-32e.html (application date 02.05.2019).

[5] Standardized ARS Interface. Technical Documentation. ARS 404-21 (Entry) ARS 408-21 (Premium). V 1.8. Continental Engineering Services GmbH, 2017.

[6] Crafter 2017. Introduction Self-Study Program 566. Commercial Vehicles 566 VW, 2017.

[7] Advanced emergency braking system / DAF Trucks Russia URL: http://www.daf.ru/ru-ru/trucks/comfort-and-safety-systems-euro-6/advanced-emergency-braking-syste (application date: 02.05.2019).

[8] Active emergency braking system // Small vehicles I Mercedes-Benz URL: https://www.mercedes-benz.ru/vans/ru/sprinter/tourer/optional-equipment-highlights-1/teaser-group-1/active-brake-assist (application date : 02.05.2019).

[9] Emergency braking - a lifesaving system // Volvo Trucks - European trucks // Volvo Trucks - URL: https://www.volvotrucks.ru/ru-ru/news/magazine-online/2017/jul/tech-focus-emergency-brake.htmlt (application date : 02.05.2019).

[10] Uniform provisions concerning the approval of motor vehicles with regard to advanced emergency braking systems // UNECE URL: http: //www.unece.org/fileadmin/DAM/trans/main/wp29/wp29regs/2015/R131r.pdf (application date: 25.08.2019).

AUTHOR INDEX

Alexeev, A.V. 115, 122
Alexeyev, A.V. 109
Alyabiev, Alexey 23
Andronov, A V 127, 133
Anisimov, P A 62
Anosov, M. S. 72
Anosov, M.S. 155
Barinov, A.A. 46
Bazhenov, E E 1
Bazhenov, M 68
Bazhenov, Y 68
Bazhenova, L V 1
Belyakov, V.V. 96, 161
Berdnikov, L.A. 109, 115
Beresnev, P O 6
Biryukov, A.L. 142
Borisov, G.V. 190
Brovkin, A.S. 122
Burov, V. N. 194
Bushmanov, D.V. 54
Chernyshev, D O 1
Chichkina, M.I. 115
Chugunov, Alexander 82
Churakov, A.V. 138
Devyatkin, A.V. 149
Erasov, I. A. 166
Erkhov, Alexander 23
Ermilov, A.E. 102
Evdokimova, T. S. 41
Fadeev, R. S. 194
Fedosova, L. O. 41
Filatov, D. M. 149
Filatov, V I 6
Filippov, A.A. 171, 177, 183
Gnenik, M.E. 161
Golovin, A.A. 155
Goncharova, D.A. 171, 177, 183
Kabaldin, Y.G. 155
Kabaldin, Yu.G. 72
Kalyashov, V.A. 138
Kapustin, A.G. 161
Khitrov, E G 127, 133
Khlybov, A.A. 155
Klubnichkin, Evgeny 16, 23
Klubnichkin, Vladislav 16, 23
Kocherov, A.V. 54
Kolesnitcenko, D.A. 194
Kolobov, M.A. 54
Korchazhkin, M.G. 109, 122
Korostelev, S A 34
Korotunova, I.A. 46, 171
Kuklina, I.G. 166
Kulikov, Roman 82
Kurochkin, S V 89
Kuzin, V.D. 77
Kuzmin, A.N. 177
Kuzmin, N.A. 171, 190
Makarov, V.S. 96, 161
Makarov, Vladimir 16
Malinin, N. V. 41
Markov, V.A. 138
Markovnina, A.I. 96
Mavleev, I.R. 190
Meshkov, A.M. 46
Mokerov, D.S. 166
Molev, J.I. 190
Molev, Y.I. 166
Moshkov, P. S. 194
Myakinkov, A. V. 194
Nuzhdin, R V 89
Okunev, A.V. 54
Pachurin, G.V. 171, 177, 183
Papunin, A.V. 161
Parfenopulo, G.K. 138
Plotnikov, S.A. 142
Podrubalov, Maxim 16
Porubov, D M 6
Pronin, P.I. 77
Rogov, P. S. 194
Salakhov, I.I. 190
Saleh, H M 89
Scriabin, M.D. 155
Shashkina, K.M. 77
Shatagin, D. A. 72
Shatagin, D.A. 155
Sidorenkov, D. A. 72
Sinodkin, A. A. 41
Sizov, A.Y. 46
Skrebneva, T.G. 46, 166
Sokolova, V.A. 138
Tiurikov, M. I. 41
Toropov, E.I. 194
Trunova, I.G. 183
Trusov, Y. P. 194
Tumasov, A V 6
Tyugin, D Yu 6
Verbilov, A F 34
Voronin, M M 62
Yarjemsky, A.D. 194
Yarzhemsky, A.D. 109, 115, 122
Yusupov, R.R. 109, 115, 122
Zamolodchikov, Vladimir 82
Zarubin, D.N. 77
Zhelonkin, M.V. 72
Zolotov, A V 62
Zubakin, A.S- 142